QUANTUM CHAOS AND QUANTUM DOTS

Mesoscopic Physics and Nanotechnology

QUANTUM CHAOS AND QUANTUM DOTS

Katsuhiro Nakamura
Takahisa Harayama

OXFORD

UNIVERSITY PRESS

OXFORD

Great Clarendon Street, Oxford OX2 6DP

It furthers the University's objective of excellence in research, scholarship, and education by publishing worldwide in

Oxford New York

Auckland Bangkok Buenos Aires Cape Town Chennai
Dar es Salaam Delhi Hong Kong Istanbul
Karachi Kolkata Kuala Lumpur Madrid Melbourne Mexico City Mumbai
Nairobi São Paulo Shanghai Taipei Tokyo Toronto

Oxford is a registered trade mark of Oxford University Press
in the UK and in certain other countries

Published in the United States
by Oxford University Press Inc., New York

© Oxford University Press 2004

The moral rights of the authors have been asserted
Database right Oxford University Press (maker)

First published 2004

All rights reserved. No part of this publication may be reproduced,
stored in a retrieval system, or transmitted, in any form or by any means,
without the prior permission in writing of Oxford University Press,
or as expressly permitted by law, or under terms agreed with the appropriate
reprographics rights organization. Enquiries concerning reproduction
outside the scope of the above should be sent to the Rights Department,
Oxford University Press, at the address above

You must not circulate this book in any other binding or cover
and you must impose this same condition on any acquirer

A catalogue record for this title is available from the British Library

Library of Congress Cataloging in Publication Data
(Data available)
ISBN 0 19 852589 3

10 9 8 7 6 5 4 3 2 1

Printed in Great Britain
on acid-free paper by
Biddles Ltd., www.biddles.co.uk

PREFACE

Chaos is by nature a word from classical mechanics which describes the macroscopic world, whereas the microscopic world obeys the laws of quantum mechanics. Chaos is therefore not a concept suitable to quantum mechanics. Roughly speaking, chaos means that future that obeys a deterministic law is probabilistic and not predictable. The unpredictability of a given system is caused not by the largeness of its degrees of freedom but by its nonlinearity. Recalling the linearity of the Schrödinger equation for microscopic systems, chaos cannot be compatible with quantum phenomena.

On the other hand, random features can show up in quantum mechanics in the case of heavy nuclei where the underlying interaction is very complicated and also in disordered and diffusive metallic systems characterized by an ensemble of impurities or by random potentials. These systems are known to be described by random matrix theory. Recently, however, a growing number of systems involving a simple interaction has turned out to have a feature consistent with the prediction by random matrix theory. It should be emphasized here that the simple interaction in these systems is rather well-defined and not random at all. The situation where random matrix theory is applicable to the system endowed with no randomness is reminiscent of the situation in classical mechanics where the motion of a particle in a system with a few degrees of freedom can be chaotic and random. Noting this similarity, one is inclined to speculate that the effect of chaos can appear in quantum phenomena. In fact, random matrix theory is fruitfully applied to many of quantized chaotic systems. Thus, we refer to the rich effects of chaos observed in quantized chaotic systems as *quantum chaos.*

Research on quantum chaos has been developed mainly around computational and theoretical themes in the fields of physics and chemistry. Among these themes of mathematical physics, one points out the study of electronic motion in the setting of billiards. Microscopic billiards are typical dynamical systems in which it is easy to see chaos classically and also to investigate the effects of chaos on quantum phenomena. Without making it into a real entity, however, the billiard system would remain a fictitious toy model.

Profound theory can find its realization. In fact, using submicron-scale semiconductor devices, a microscopic billiard table called the *quantum dot* came into existence recently. Owing to the progress in microstructure fabrication technology, a microscopic billiard table has been made by processing the interface layer at a compound-semiconductor heterojunction, and is nowadays in full flourish as a system in which to study quantum chaos as natural science rather than as a sophisticated theory.

In this book, we shall explain quantum chaos and quantum transport in microscopic billiards. Although research on quantum chaos has so far been developed in wider directions, we shall limit the scope to billiard systems, which are imaginable, being easily understood, and therefore very convenient for beginners. If, before access to quantum chaos, readers are to be involved in the intensive study of the features of classical mechanics for general Hamiltonian systems, most of them would feel bored and would give up proceeding towards even an introduction to the subject of quantum chaos. In fact, to understand quantum chaos, it is not always necessary for us to be armed with comprehensive knowledge of classical chaos. Consequently, in this book, only a minimum introduction to the nature of classical chaos is provided, which is necessary to learn quantum chaos in microscopic billiards. Throughout the text, however, we choose the *semiclassical approach*, and emphasize the role of the *semiclassical theory* which makes it possible to interpret quantum phenomena in the language of the classical world. This means that we keep the spirit of Bohr's correspondence principle by trying to explore the connection with the underlying classical *nonlinear dynamics*. Consequently, not much attention is paid to issues directly obtained by random matrix theory.

While the subject covered here is confined to ballistic microstructures like quantum dots and antidots, there are other essential themes of quantum chaos that are suppressed in this book. We must ask for readers' understanding about this omission. Nevertheless, the way of thinking, the analytic method, and the approximations explained in this book are prototypes in the study of quantum chaos and it is straightforward to apply the scheme to generic systems other than billiards. We are naturally convinced that the material presented here would be extremely beneficial to students and researchers wishing to become acquainted with quantum chaos.

In Chapters 1 through 4, we describe what is observed in the current research and where the interest is concentrated, and explain the typical and basic theory of quantum chaos in quantum transport. Special care is taken so that one can read these chapters without having recourse to other technical books. Chapters 5 through 10 are concerned with the application of the basic theories, dealing with the topical subjects arising from the crossing between quantum chaos and quantum transport. In particular, we employ concrete themes in mesoscopic quantum transport in quantum dots – quantum interference, ballistic weak localization, together with the criticisms against it, persistent current, orbital diamagnetism, fractal magneto-conductance fluctuations, Arnold diffusion, Coulomb blockade, etc. All these are rapidly progressing in both experiment and theory, and, by resorting to semiclassical theory, we try to elucidate their relationship with deterministic chaos. After reading this part, readers will be able to get started with research on quantum chaos in quantum dots.

While several technical books on quantum chaos already exist, it is not easy to write an introductory book for beginners that also incorporates topical subjects in this field. This book stems from our Japanese book entitled "*Quantum Chaos - On the Stage of Quantum Billiards*" (Baifukan Publishing Co., Tokyo, 2000). But, supplemented by a lot of updated material, it has become a thoroughly revised and greatly extended version of the original. Also we have rewritten as plainly as possible for the present edition. Consequently this introductory book on quantum chaos and quantum dots explains the subject step by step in detail from the very beginning through to applications. The authors hope this book will attract a wide audience to this important field full of scientific romanticism.

We have to thank a lot of people for their generous contributions. K.N. is very grateful to Prof. Michael Berry at the University of Bristol for his consistent encouragement and support. T.H. expresses many thanks to Dr. Bokuji Komiyama, Director of ATR Adaptive Communications Research Laboratories at Kyoto for his continued encouragement and stimulating discussions. Both the authors have benefited from the following people who have inspired and given us innumerable suggestions and comments: Y. Aizawa, T. Aida, J. Bird, O. Bohigas, G. Casati, P. Davis, N. Egami, T. Fromhold, P. Gaspard, T. Geisel, G. Hackenbroich, H. Hasegawa, M. Hosoda, K. Ikeda, S. Iwabuchi, T. Iyetake, T. Kato, J. Keating, R. Ketzmerick, H. Makino, M. Nakayama, E. Narimanov, Y. Ochiai, T. Prosen, L. Reichl, K. Richter, Y. Liu, S. Saito, M. Sano, S. Sawada, T. Simizu, M. Stopa, A. Terai, H. Weidenmüller, D. Weiss, and M. Wilkinson (in alphabetical order). We are particularly indebted to our most recent collaborators: A. Budiyono, S. Kawabata, J. Ma, A. Shudo, Y. Takane, and S. Tasaki. They contributed immensely to the work and to the physical understanding of the subject. Special thanks go to Prof. Izumi Kubo who told us about Kelvin's original work on billiards and to Prof. Muthusamy Lakshmanan for his critical reading of the manuscript and correcting its grammatical errors.

September, 2003

Katsuhiro Nakamura

Takahisa Harayama

CONTENTS

1

QUANTUM CHAOS AND BILLIARDS

Chaos is a macroscopic phenomenon which obeys classical mechanics. Nonlinearity in classical mechanics is essential for us to see chaotic phenomena. We cannot expect to observe chaos in microscopic phenomena ruled by quantum mechanics, however, since the Schrödinger equation describing the wavefunction or probability amplitude is strictly linear. Nevertheless, quantum manifestations of chaos in classical mechanics have been predicted theoretically and numerically[1-29], and observed in actual experiments for the last few decades[30]. In this chapter, after addressing the historical background of this field, the notion of quantum chaos will be introduced by giving prototypical examples emanating from the experimental results of quantum transport in mesoscopic semiconductor devices.

1.1 Birth of the Physics of Billiards

The dynamics of billiard balls and its role in physics have received wide attention since the monumental lecture by Lord Kelvin (the man whose name is used as the unit of absolute temperature) at the turn of the 19th century. We shall begin with a brief review of his lecture[31].

On Friday 27 April 1900, at the Royal Institution of Great Britain, he delivered a lecture entitled "The 19th century clouds over the dynamical theory of heat and light." The first cloud was the question of the existence of the **ether** propagating light. He disapproved of the possibility of the Earth moving through the ether. The second cloud was the question of the validity of the Maxwell–Boltzmann (MB) distribution leading to the equipartition of energy and he ultimately doubted the **ergodicity hypothesis** behind the MB distribution. Five years later after Kelvin's lecture, the first cloud was swept away by Einstein's special theory of relativity. But, how did the second cloud disappear?

The ergodicity hypothesis is the assumption that a long time average of a given physical quantity should accord with its phase-space average. Choose as an example an ideal gas consisting of noninteracting atoms with no internal degrees of freedom, and define the ratio of its two specific heats (at constant pressure and at constant volume). By paying attention to the low-density and high-temperature region where neither quantum effects nor interatomic interaction is present, Kelvin addressed a small but finite discrepancy of the above ratio between the theoretical issue predicted by the equipartition of energy and the experimentally observed value. Noting further that this discrepancy is enhanced in molecules with rotational degrees of freedom as well as translational ones, he insisted on a breakdown of the ergodicity ansatz.

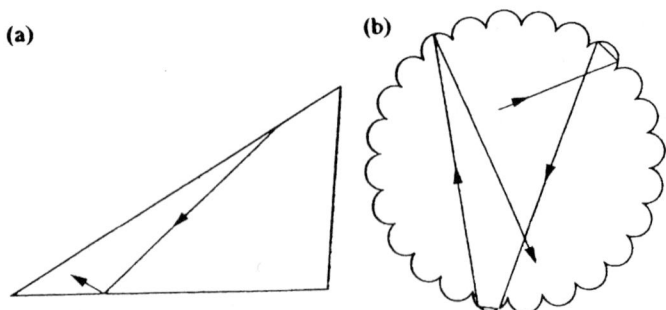

Fig. 1.1   Kelvin's billiards: (a) triangle; (b) flower.

To demonstrate more explicitly the breakdown of the ergodicity hypothesis, Kelvin (with his assistant, Anderson) investigated point-particle motion bouncing off the hard wall of a triangular billiard table (see Fig. 1.1(a)). Measuring each line segment between successive bounces and each reflection angle at the wall repeatedly (600 times!), he showed the breakdown of the equipartition of energy, i.e., the inequivalence between long-time averages of transverse and perpendicular components of the kinetic energy. Next, he chose the flower-like billiard table in Fig. 1.1(b), carried out a similar pursuit, and again showed the long-time averages of the radial and angular parts of the kinetic energy not to satisfy the equipartition of energy. This investigation led to the birth of the physics of billiards. The physics of billiards was thus launched on 27 April 1900.

While the triangular billiard system employed by Kelvin is pseudo-integrable according to the modern terminology of mathematics, the flower-like system is nonintegrable, exhibiting chaos as proven much later by Bunimovich. Strictly speaking, in his investigation above, Kelvin approximated the flower-like billiard table with a polygonal one, by replacing each small arc of the flower by its chord, and had recourse to playing cards to determine the position on each chord at which a point particle bounces. He dared to choose this simplification because a point particle shows complicated multiple scattering inside the arc each time it enters the arc, making the precise measurement of line segments very difficult. This simplification, however, was the principal reason to break the ergodicity hypothesis.

Hence, in order to sweep away the 19th century cloud over the ergodicity hypothesis, it had become indispensable to envisage complex features of nonlinear dynamics of a particle in billiards with the shape like in Fig. 1.1(b). In fact, an accumulation of studies on billiards (by Birkhoff, Krylov, Sinai and others[32-36]) during the 20th century since Kelvin's lecture was devoted to those on nonintegrable and chaotic billiards. Among them, the **stadium billiard** table composed of a rectangle attached to a pair of semicircles is the most typical, with the billiard ball moving inside the hard wall of the stadium. The dynamics of the ball is chaotic, i.e., extremely sensitive to initial conditions: a very slight variation in initial coordinates or momenta yields a thoroughly different orbit. The sensitivity to initial conditions causes a cluster of initial points with similar initial conditions (comparable to a droplet of cream fallen in coffee) to exhibit mixing in phase space as time elapses just as the cream is mixed in coffee, and thereby to show an ergodic property. In this way, chaotic billiards like a stadium have resolved the long-standing problem (the second cloud of Kelvin), and constitute cornerstones by which to consolidate the foundations of statistical mechanics.

Since the ground-breaking experiments in the 1990s[37-42], the physics of billiards has developed in every direction of science and technology. Billiards can nowadays be created as **quantum dots** in the microscopic world, and one can envisage quantum-mechanical manifestations of the chaos of billiard balls. In fact, owing to recent progress in advanced technology, submicron-scale quantum dots such as chaotic stadium and regular circle billiards are fabricated at the interface of semiconductor heterojunctions. Here, the electron plays the role of a billiard ball, and shows deterministic **ballistic motion** thanks to having a dot size less than the mean-free path of the electrons. Puzzling experiments on resistance fluctuation for stadium and circle dots are motivating the exploration of the effect of billiard-ball dynamics on ballistic quantum transport.

1.2   What is Quantum Chaos?

On the other hand, the study of **chaos** in general from a mathematical point of view also has a long history and is traced back to Poincaré's ideas at the end of the 19th century. Owing to dramatic progress in computer science since the 1960s, everybody working in any field of science has nowadays come to recognize the great significance of chaos.

In fact, we often encounter chaos, not only in natural phenomena such as the flow of a river and motions of clouds or planets, but also in games such as roulette and dice. What is the common property that the above phenomena possess? First, they are macroscopic, and have no uncertainty. Accordingly, the systems' states in the very near future can be predicted precisely, that is, these systems are deterministic. Second, after a very long time the states cannot be predicted because their time evolution is unstable and looks random. Thus, chaos has both deterministic and random natures, which might sound contradictory. The reason why the randomness appears is that an extremely small error in the initial condition which one cannot identify grows exponentially in the course of the time evolution due to the nonlinearity of the systems, and finally results in a thoroughly different state.

The word **quantum chaos** might remind us of unpredictable behavior in quantum phenomena, but this is not the truth. The solution of the linear Schrödinger equation cannot behave chaotically in the same way as that of the Newtonian equation.

Quantum chaos means a **quantum manifestation of chaos in deterministic classical mechanics**. Here, the manifestation of chaos is the common characteristic phenomena of quantized classically chaotic systems. First we describe the system within classical mechanics to find chaotic aspects, describe it within quantum mechanics, and then find the effects of chaos. But, is it really possible to relate phenomena in classical mechanics to those in quantum mechanics? Basically, classical mechanics which describes macroscopic phenomena is totally different from quantum mechanics for microscopic phenomena. But, let us remember **Bohr's correspondence principle** which claims that the descriptions by classical and quantum mechanics should give the same result when they describe the same **macroscopic system**. Accordingly, when the energy of the microscopic system is sufficiently high, we can expect to find the quantum-mechanical effect of chaos in classical mechanics. From this viewpoint, it seems inevitable to learn about chaos in classical mechanics before understanding quantum chaos. Therefore, the textbooks of quantum chaos conventionally start from a long review of chaos in classical mechanics. However, this is a difficult way for researchers or students to learn quantum chaos. Here we shall take an alternative route.

In Chapter 2, after a brief introduction to chaos and quantum mechanics, we shall sketch the theory of quantum chaos which roughly explains the experiment in Chapter 1. Chapters 3 and 4 deal with the fundamental theory of quantum chaos. A detailed description of chaos in classical dynamical systems will be given in Chapter 3. The experiment in Chapter 1 will be explained in detail from the viewpoint of quantum mechanics in Chapter 4.

Chapters 5 through 10 will be concerned with applications of the fundamental theory of quantum transport in a ballistic regime where the deterministic law is operative, and offer a lot of novel examples of the manifestation of chaos that will show up in weak localization, Aharonov–Bohm oscillations, orbital diamagnetism in quantum dots and quantum transport in antidot lattices, etc. In this book, our efforts are also devoted to help readers to understand quantum chaos as concretely as possible. Therefore, we focus on the topics of quantum transport phenomena, and explain the fundamental framework and methodology of quantum chaos by always having recourse to the experiments of semiconductor devices.

1.3 Resistance of Quantum Dots

Figure 1.2 shows the result of an experiment for two quantum-effect devices (quantum dots). The vertical and horizontal axes denote the resistance and the strength of magnetic field B, respectively. The sensitive dependence of the resistance on magnetic field implies the resonance scattering as a quantum effect, because the size of devices is so small as to be on the submicron (10^{-7}m) scale. The inset of Fig. 1.2 is a magnification of the region close to zero magnetic field.

Let us pay attention to the difference between the two sets of data. One can easily find that the zero-field peak (the peak in the vicinity of zero field) in Fig. 1.2(b) is much steeper than that of Fig. 1.2(a).[1] An intensive analysis of this particular peak will be made in Chapter 7. In the following, we try to extract the characteristics of

---

[1] According to later experiments, the peak profile is more or less affected by some subtle averaging. However, the essential part of the data in Fig. 1.2 remains unchanged. See more details in Chapter 7.

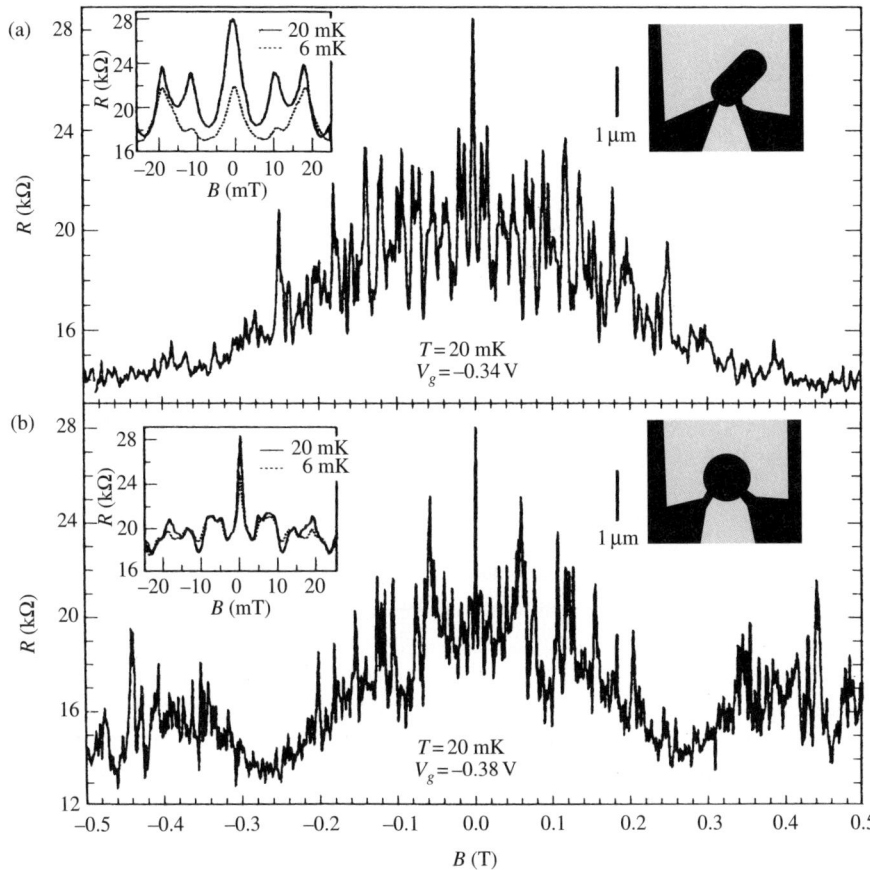

Fig. 1.2 Resistance versus magnetic field: (a) stadium; (b) circle (reprinted from[37]).

the whole data of Fig. 1.2 by processing several transformations.

The statistical properties of the **conductance** g, defined by the inverse of resistance, are also very important when the original data show a violent oscillation with respect to magnetic field, which can convey a **universal aspect** of devices as explained afterwards. Therefore, dividing the conductance into the mean part $\bar{g}$ and fluctuations around it, δg, we shall investigate the **statistical property of fluctuations** of conductance by analyzing the **autocorrelation function** $C(\Delta B)$ defined by

$$C(\Delta B) \equiv \langle \delta g(B + \Delta B)\delta g(B)\rangle_B \qquad (1.1)$$

where $\langle \cdots \rangle_B$ denotes the average over the magnetic field B. The **power spectrum** is defined by the Fourier transformation of the autocorrelation function, and it ex-

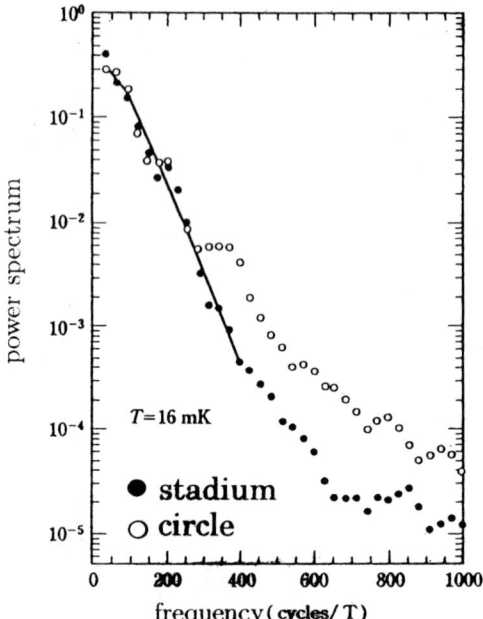

Fig. 1.3   Power spectra of conductance fluctuations (reprinted from[37]).

tracts the characteristics of the frequency distribution included in the autocorrelation function. The autocorrelation function and the power spectrum are shown in Fig. 1.3. Filled and white circles correspond to Fig. 1.2(a) and (b), respectively. In Fig. 1.3 there is a clear difference especially in the region whose frequencies are higher than 250 cycles/T.

Why do the data from these two devices differ from each other? Although different devices show different data, there is not as much difference as the number of devices. In fact we can classify the types of these devices into two categories by measuring their resistance. The difference between the devices can be recognized in the shapes of the structure located near their center, the stadium in Fig. 1.2(a) and circle in Fig. 1.2(b).

As a result, one may summarize: the magnetic-field dependence of resistance is controlled by the shape of devices, and, according to the power spectra of their conductance fluctuations, their shapes can be classified into two distinct categories.[2]

Before proceeding to this kind of analysis, we shall explain how to manufacture the devices under consideration. The devices are fabricated at the interface layer of a GaAs/AlGaAs compound-semiconductor heterojunction. Along the interface, a

---

[2]In this statement, we limit the device shapes to those that admit "completely chaotic" or "completely integrable" motion of an electron. In devices with a generic shape, however, both chaotic and regular motions can coexist. For details, see Chapters 7 and 9.

triangular potential-well is formed to which the motion of carrier electrons is limited, resulting in the two-dimensional conducting region. By using additional fabrication of this region, the two-dimensional (2-d) electron gas is finally confined to a finite planar area of size less than $1\,\mu$m, which is called a **quantum dot**. The shape of the quantum dot is a stadium in Fig. 1.2(a) and a circle in Fig. 1.2(b). A nearly one-dimensional device with a width of $0.1\,\mu$m can also be fabricated, and this is called a **quantum wire**. In Fig. 1.2 a pair of quantum wires as leads is attached to each quantum dot. Electrons in the quantum dot are coming from and going to leads. Since the size of the device is on the submicron scale and the temperature is very low, the mean free path is much longer than the length scale of the devices. At low temperatures $\sim 20\,$mK, the mean free path of electrons is about $2.6\,\mu$m and is much larger than the scale of both quantum dots and wires. In this situation electrons show **ballistic** rather than diffusive motions, namely, obey deterministic laws. A quantum dot attatched to quantum wires as leads constitutes an open billiard system (see Fig. 1.2) for which ballistic quantum transport is intensively investigated both experimentally and theoretically.

1.4 Dynamics in Billiards and Semiclassical Theory

According to the difference in the features of resistance, quantum dots can be classified into two categories as mentioned in the last section. Various shapes of dots are depicted in Fig. 1.4. The shapes on the left (right) panel bear the same feature of the resistance as the stadium (circle).

Amazingly, this classification accords with one predicted on the basis of a completely different viewpoint. The new criterion is a **dynamical feature of billiards, i.e., nonlinear dynamics of a billiard ball**. Suppose one is to play a billiard game using a billiard table of various shapes. Here we assume the idealized situation: a billiard ball is taken as a point particle and the friction between the ball and billiard table is ignored. According to the nature of motion of the billiard ball, the shapes of the billiard tables can be classified in the same way as in Fig. 1.4. The balls on the billiard tables with shapes in Fig. 1.4(b) show a regular motion. For instance, the ball on the circle executes a regular trajectory in Fig. 1.5(b). The motion on any other billiard table depicted in Fig. 1.4(b) is known to be regular as well.

On the contrary, the motion on the stadium-shaped billiard table is quite irregular as seen in Fig. 1.5(a). This **irregular and unpredictable motion is called chaos** and the motion on any billiard table in Fig. 1.4(a) is known to be chaotic. In this way, the shapes of quantum dots can be classified according to the feature of the billiard motion, and this classification exactly agrees with the one based on the feature of resistance in quantum dots.

However, one would become aware of a great gap that is not being filled: the quantum dot is a quantum-mechanical object, whereas the billiard system is a classical one. Resistance is a quantum-mechanical phenomenon, but the motion on a billiard table is a phenomenon that obeys a deterministic classical law. The most serious question to be addressed is why quantum dots and classical billiards are classified in

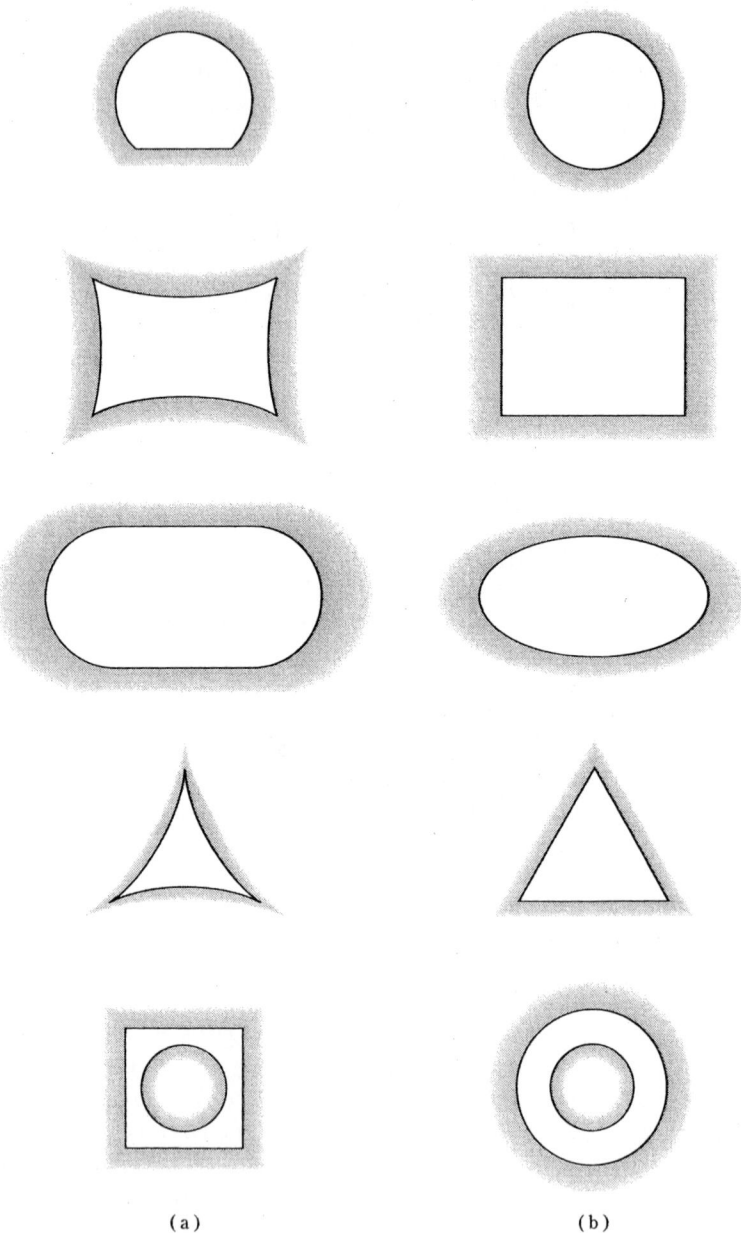

(a) (b)

Fig. 1.4 (a) Billiards leading to chaos; (b) billiards leading to regular motion.

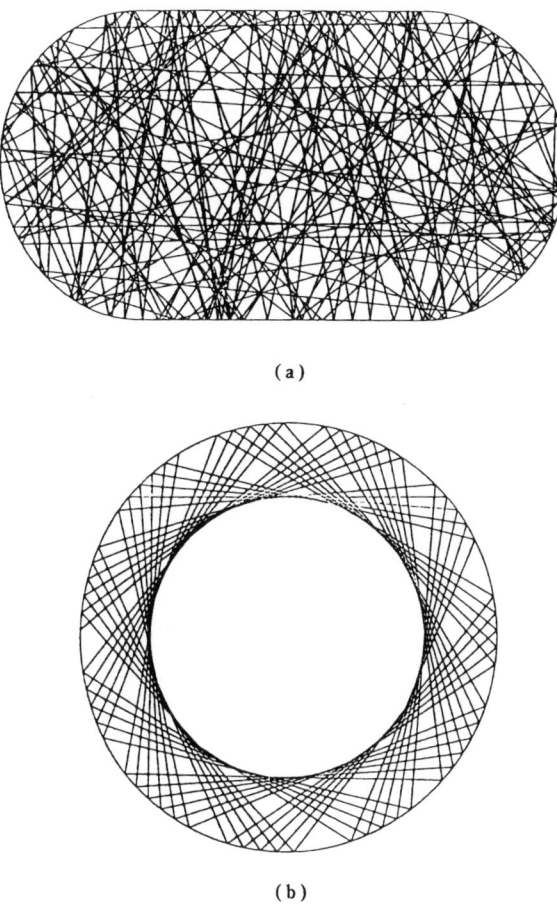

(a)

(b)

Fig. 1.5   Motion of billiard ball: (a) chaos; (b) regular motion.

a similar way according to their shapes despite the difference of the worlds in which they exist, microscopic and quantum mechanical versus macroscopic and classical mechanical.

It is precisely the main purpose of this book to give a clear answer to this question and to capture the influence of classical chaos on quantum phenomena. But is it really possible to connect two entities that appear not to be related at all?

The distinction between the microscopic and quantum-mechanical world and the macroscopic and classical-mechanical world lies in the scale of the system's energy. With increase of the system's energy in the microscopic system, the quantal and classical-mechanical results should become identical. In other words, the quantum-mechanical description of a macroscopic phenomenon should accord with the result of the classical description. This is just a notion of **Bohr's correspondence prin-**

ciple. The system's energy would be a key to interpreting the quantum world in the language of the classical world. To be explicit, in treating the quantum theory of resistance, one can regard the energy (or precisely speaking, the action) of the electron to be much larger than the Planck constant and may employ the approximation that ignores the corrections in higher orders of the Planck constant. This is the **semiclassical approximation** and the theory that describes the quantum phenomena with the use of the semiclassical approximation is called the **semiclassical theory**. The semiclassical theory serves as a dictionary in our translating the language of the classical world into that of the quantum one, and is also a powerful tool, wherever the de Broglie wavelength is much less than the system's length scale. By applying the semiclassical theory to the resistivity analysis of quantum dots, the reason would become transparent as to why the features of the motion of billiard balls is responsible for the features of quantum transport and thereby one can recognize why the classification of quantum dots via resistivity fluctuations is identical to that of billiards via the dynamics of billiard balls. Since we try to keep to the spirit of Bohr's correspondence principle, major attention will be paid to the semiclassical theory rather than to random matrix theory.

In Chapter 2, we shall sketch a general framework of applying the semiclassical theory to the analysis of quantum transport in quantum dots. More detailed description will be given in Chapters 3 and 4.

2

QUANTUM TRANSPORT AND CHAOS IN BILLIARDS

At first sight, electrical conductance in a microscopic quantum dot might seem to have nothing to do with the motion of a billiard ball. The experimental evidence mentioned in Chapter 1, however, indicates a strong correlation between these two events belonging to different worlds, and the semiclassical theory provides a powerful tool to shed light on such a correlation. Thanks to the semiclassical theory, quantum phenomena can be described and understood in terms of a knowledge of classical mechanics. In this chapter, by applying the semiclassical theory to electrical conductance in a quantum dot, we shall come to recognize how the conductance fluctuations are related to billiard-ball dynamics.

2.1 Quantum Theory of Conductance

Let us start with a planar quantum dot with 2-d lead wires as shown in Fig. 2.1. How does quantum mechanics treat the quantum phenomenon of electrical conductance through the dot? The attached lead wires are rectangular and semi-infinite with their width finite in the direction perpendicular (transverse) to the leads. Due to the transverse confinement, the eigenstates of an electron in the leads are finite in number and their eigenenergies are discrete.

The rate of transition of an electron in one state at lead 1 to another at lead 2 via transmission through a quantum dot determines the conductance. For the quantum dot in Fig. 2.1, the conductance g is given by the **Landauer formula**[43] (its generalization to the multilead case is called the **Landauer–Büttiker formula**):

$$g = \frac{2e^2}{h} \sum_{n=1}^{N_M} \sum_{m=1}^{N_M} |t_{nm}|^2 \tag{2.1}$$

Fig. 2.1   Quantum dot with lead wires.

with electronic charge e and Planck constant h. The prefactor 2 is due to the spin degree of freedom of the electron. m and n are **mode numbers** related to the state of an incident electron from lead 1 and that of an outgoing electron through lead 2, respectively. The mode number is nothing but the number of half wavelengths for the standing wave which is the perpendicular component of the wavefunction at each lead. N_M is the upper limit of m and n, denoting the possible number of propagating modes. $|t_{nm}|^2$ is the **transmission probability** that an electron starting from mode m at lead 1 will reach mode n at lead 2.

t_{nm} itself is the transmission amplitude that the electron arriving at the dot (cavity) in mode m propagates inside the dot through the Green function and finally exits in mode n. To be explicit, it is given by[44]

$$t_{nm} = -i\hbar(v_m v_n)^{1/2} \int dy' \int dy \phi_n^*(y') \phi_m(y) G(L, y'; 0, y; E), \qquad (2.2)$$

where $\phi_m(y)$ and v_m are, respectively, the wavefunction in the transverse (y) direction and the parallel component of velocity vector at lead 1, and similarly for $\phi_n(y')$ and v_n at lead 2. (Electrons contributing to the electrical conductance are limited to those in the vicinity of the Fermi sea, so the energy E denotes the Fermi energy E_{F}.) $G(0, y'; L, y; E)$ is the (retarded) **Green function** for the electronic propagation from a point $(x = 0, y)$ of contact of the quantum dot with lead 1 and to another one $(x = L, y')$ with lead 2.

The above scenario is purely quantum-mechanical. So long as one's interest lies in calculation of conductance, quantum mechanics is a perfect tool to accomplish this purpose, and one needs no extra theory like the semiclassical one. However, our purpose is not limited to the computation of conductance. Rather, it is the central theme of this book to capture the effects of classical chaos that would show up in conductance. So we should be familiar with the semiclassical theory, namely, a way to apply the knowledge of chaos to (2.2). Another important aspect of the semiclassical theory is that it makes it possible to evaluate the conductance analytically rather than numerically. In fact, for a quantum dot with the shape depicted as in Fig. 1.4(a), one has no analytical expression for the Green function. Nevertheless, the semiclassical theory can provide an approximate expression for the Green function and can eventually derive an approximately analytic formula for conductance fluctuations.

2.2   Semiclassical Approximation and Stationary-Phase Method

The Green function and transmission coefficient are functions of the action and the Planck constant ($\hbar$). Since the action itself grows with the system's energy, one may intuitively regard a larger energy as a larger action. In the context of quantum transport through submicron-scale quantum dots, the kinetic energy of the electron is sufficiently high with its action much larger than ($\hbar$). Let us explain the **stationary-phase approximation** justified in this limit.

Consider the following integral

$$I = \int_{-\infty}^{\infty} f(x) \exp\left(i\frac{\phi(x)}{\hbar}\right) dx, \tag{2.3}$$

where $f(x)$ is a smoothly varying function of x and the derivative of $\phi(x)$ is assumed to vanish only at $x = 0$. In the path-integral representation of the Green function in quantum mechanics, the action plays the role of $\phi(x)$. In the limit $\phi(x) \gg \hbar$, the phase factor in the integrand in (2.3) is extremely sensitive to the variation of x. This fact together with the notion of smoothness of $f(x)$ gives rise to a violent undulation of the integrand in (2.3), whose integral should vanish in most of the integral zone. An exception occurs near $x = 0$ where $\phi(x)$ is stationary, leaving a finite integrated value. Therefore the integration limited to the vicinity of $x = 0$ where the phase is stationary should be a good approximation for I in (2.3), from which the terminology of stationary-phase approximation stems. Near the stationary point, $\phi'(0) = 0$ and

$$\phi(x) \approx \phi(0) + \frac{1}{2}\phi''(0)x^2. \tag{2.4}$$

Using (2.4) in (2.3), we have

$$I \approx f(0) \exp\left(i\frac{\phi(0)}{\hbar}\right) \int_{-\infty}^{\infty} \exp\left(i\frac{\phi''(0)}{2\hbar}x^2\right) dx, \tag{2.5}$$

which reduces to

$$I \approx f(0)\sqrt{\frac{2\pi\hbar}{|\phi''(0)|}} \exp\left\{i\frac{\phi(0)}{\hbar} + \frac{\pi}{4}i\,\mathrm{sgn}\,\phi''(0)\right\}. \tag{2.6}$$

Here $\mathrm{sgn}(a)$ means the sign of a. In general cases that have several isolated stationary points other than $x = 0$, the right-hand side of (2.6) is replaced by the sum of contributions of a similar kind.

To summarize, the stationary-phase approximation, which works well for an integral of the integrand with its phase factor extremely sensitive to the integration variable x, gives an approximate evaluation of the integral by expanding the phase around the stationary-phase point up to quadratic in x. Below we shall illustrate how this approximation is quite beneficial in bridging quantum mechanics and classical mechanics.

2.3 Semiclassical Green Function and Transmission Amplitude

The way to apply the semiclassical theory to the Green function in a quantum dot differs between the shapes in Fig. 1.4(a) and Fig. 1.4(b). Here we concentrate on the case of Fig. 1.4(a).

The Green function describes the amplitude of propagation of a quantum particle. According to Feynman's path-integral method, one considers an infinitely large number of possible paths starting at one point and terminating at another and assigns to each path an exponential factor with its phase given by the path-dependent action (integration of momentum along each path). Then the Green function is expressed as the sum of the exponential terms over all these paths. Therefore originally the Green function accommodates no picture of classical paths. In other words, for a quantum-mechanical billiard ball, there is no need to obey the law of reflection (equivalence between angles of incidence and reflection) on its collision with the wall. On the other hand, in the classical limit, a billiard ball is governed by Newton's laws with suitable boundary conditions at leads 1 and 2, and only paths that satisfy the **law of reflection** are accepted to contribute to the transmission. In the semiclassical border between quantum and classical mechanics, both the classically admissible paths and the effect of paths slightly deviating from them are taken into consideration: in Feynman's path-integral representation[3] of the Green function for a particle with sufficiently high energy, only the paths that satisfy the condition for a stationary phase can give a nonvanishing contribution, and this condition is nothing but the **variational principle** in classical mechanics. So the stationary-phase approximation is alternatively called the **semiclassical approximation**.

To state concretely in the limit that the action is much larger than the Planck constant, one obtains the Green function in the semiclassical approximation:

$$G(L, y'; 0, y; E) \approx \sum_{\text{Orbit}:\gamma} w_\gamma \exp\left(\frac{i}{\hbar} S_\gamma\right), \tag{2.7}$$

where $\sum_{\text{Orbit}:\gamma}$ implies summation over all orbits starting at $(0, y)$ and finally reaching (L, y') after bouncing with the billiard table wall. The bouncing of a ball off the wall is perfectly elastic, satisfying the law of reflection. w_γ is a quantity characterizing how each orbit γ deviates from its original one against a slight change of the initial condition. The action S_γ is the integration of momentum along the trajectory of a given orbit γ. Thus, the Green function for a quantum dot is expressed as a summation over paths of a ball on the billiard table with its shape similar to the dot. This is the first step to capture the quantum world in the language of the classical world.

Substitution of (2.7) into (2.2) together with an additional use of the semiclassical approximation leads to the transmission amplitude expressed as a sum over all paths of the billiard ball:

$$t_{nm} \approx \sum_{s(n,m)} d_s \exp\left(\frac{i}{\hbar} \widetilde{S}_s\right). \tag{2.8}$$

In the paths appearing in (2.8), there is no constraint on the initial (y) and final (y') coordinates perpendicular to the leads. Instead, the angle of incidence to the dot and that of ejection from the dot have values specified by m and n, namely, by the mode numbers at leads 1 and 2, respectively. $\widetilde{S}_s$ and d_s is the effective action and the stability factor, respectively, for the orbit s available from the billiard-ball dynamics.

2.4 Autocorrelation Function

The semiclassical formula for conductance is obtained by combining the semiclassical expression for the transmission amplitude (2.8) with the Landauer formula (2.1) (Jalabert *et al.*[45-47]). For an incident electron with wave number k, let us define the dimensionless conductance $T(k) \equiv \frac{h}{2e^2} g(k)$ which is called the transmission coefficient. Its semiclassical expression is given by

$$T(k) = \sum_{n=1}^{N_M} \sum_{m=1}^{N_M} T_{nm}$$

$$= \sum_{n=1}^{N_M} \sum_{m=1}^{N_M} |t_{nm}|^2 = \sum_{n=1}^{N_M} \sum_{m=1}^{N_M} \sum_s \sum_u f_{n,m}^{s,u}(k) \tag{2.9}$$

with N_M the total number of propagating modes at each of the leads. In (2.9), $f_{n,m}^{s,u}(k)(\equiv d_s d_u{}^* \exp(\frac{i}{\hbar}(\tilde{S}_s - \tilde{S}_u)))$, which comes out from the cross-product of t_{nm} in (2.8) and its complex conjugate, indicates the quantum inteference between orbits s and u.

The **autocorrelation function** (1.1) describing the conductance fluctuation as a function of the increment of the magnetic field, ΔB, is now written in the dimensionless form as

$$C(\Delta B) \equiv \langle \delta T(k, B + \Delta B) \delta T(k, B) \rangle_k, \tag{2.10}$$

where $\delta T(k, B)(\equiv T(k, B) - \overline{T(k, B)})$ is a **fluctuation around the mean** $\overline{T(k, B)} = \langle T(k, B) \rangle_k$. $\langle \cdots \rangle_k$ denotes the **average over wave number** (or energy), which stands for the average over the magnetic field B thanks to the system's ergodicity.

From the semiclassical formula (2.9), one recognizes that the mean transmission coefficient $\overline{T(k, B)}$ is determined by the terms with a pair of identical orbits with $s = u$, and corresponds to completely classical-mechanical transmission; by contrast, the fluctuating part $\delta T(k, B)$ is governed by quantum interference terms with $s \neq u$. From its definition in (2.10), the autocorrelation function $C(\Delta B)$ is a product of two transmission coefficients with each determined by a combination of a pair of orbits. As a consequence, four orbits will join each term contributing to $C(\Delta B)$. Assigning t and v to two additional orbits, the phase part of each term is thus written as

$$\tilde{S}_s(B + \Delta B) - \tilde{S}_u(B + \Delta B) + \tilde{S}_t(B) - \tilde{S}_v(B), \tag{2.11}$$

where $s \neq u$ and $v \neq t$ is assumed because of the suppression of the classical transmission.

2.5 Conductance and Area Distribution

The total phase (2.11) is a combination of four paths (s, u, t, and v). As the system's energy grows, this phase scaled by the Planck constant becomes very large, and, when exponentiated, shows a violent oscillation as a function of energy. Since (2.10)

assesses the averaging over wave number or energy, only the combination which makes
(2.11) almost vanish can contribute to (2.10). This particular combination is the one
satisfying $v = s$ and $t = u$ at the same time. Other possibilities $s = u$ or $v = t$ should
be excluded since $\delta T(k, B)$ is defined by suppressing the mean $\overline{T(k, B)}$.

On the other hand, the action $\widetilde{S}_s$ for a given path s under the magnetic field B is
slightly different from that under $B + \Delta B$. $\widetilde{S}_s$ is given by an integral of momentum
along the path s. In the presence of the B-field, the momentum acquires an additional
term proportional to the vector potential, which, when integrated along the path, gives
rise to the area integral of the B-field (à la Stokes' theorem). Therefore the total phase
(2.11) that survives consists of

$$\widetilde{S}_s(B + \Delta B) - \widetilde{S}_s(B) = \frac{\hbar \Delta B}{\Phi_0} \Theta_s \tag{2.12}$$

for the path $s(= v)$ *minus* its counterpart for $u(= t)$. In (2.12), $\Phi_0 \equiv \dfrac{hc}{e}$ is the **flux
quantum** and Θ_s is the area $(\times 2\pi)$ enclosed by the orbit s depicted by an incident
electron until it escapes from a quantum dot. After all, we have

$$C(\Delta B) = \frac{1}{4} \int_{-1}^{1} d(\sin\theta) \int_{-1}^{1} d(\sin\theta')$$
$$\times \sum_{s(\theta,\theta')} \sum_{u(\theta,\theta')(u \neq s)} \widetilde{A}_s \widetilde{A}_u \exp\left[i\left(\Theta_s - \Theta_u\right)\frac{\Delta B}{\Phi_0}\right] \tag{2.13}$$

where the amplitude $\widetilde{A}_s$ characterizes the deviation of an orbit s at the terminal lead
2 against a small change in the initial condition. θ and θ' are the angle of injection
from lead 1 to a quantum dot and that of ejection from the dot to lead 2, respectively.
Supposing that mode numbers at leads are large enough, the discrete sum $\sum\limits_{n=1}^{N_M} \sum\limits_{m=1}^{N_M}$ in
(2.9) has been replaced by the integral over θ and θ' in (2.13).

Noting the billiard-ball motion to be chaotic, we can proceed with the calculation. Although a detailed derivation will be given in Chapter 4, the inverse Fourier
transformation of $C(\Delta B)$ in (2.13), namely, the **power spectrum** of conductance
fluctuations, is eventually expressed in the form of convolution:

$$\widehat{C}_D(\eta) = \frac{\Phi_0 \pi}{2} \int_{-\infty}^{\infty} d\Theta P(\Theta + \eta) P(\Theta), \tag{2.14}$$

where $P(\Theta)$ is a distribution function for the area Θ enclosed by an incident orbit
until it leaves the quantum dot.

One comes to realize that the power spectrum of the autocorrelation function for
conductance fluctuations can be expressed as a convolution of the area distribution
$P(\Theta)$ for a billiard-ball orbit. In other words, quantum-mechanical phenomena characterized by conductance for a quantum dot can be captured through the thoroughly
classical information on billiard-ball dynamics.

The remaining problem is to establish the **area distribution** $P(\Theta)$ by investigating the billiard-ball dynamics within classical mechanics. To mention the conclusion in advance, in the class of billiards in Fig. 1.4(a) where the ball's motion is fully chaotic, $P(\Theta)$ always obeys an **exponential law**,

$$P(\Theta) \propto \exp(-\alpha_{\mathrm{cl}}|\Theta|) \tag{2.15}$$

with the constant α_{cl} proper to individual billiards.

By the way, how is (2.15) obtained? Suppose that an incident ball bounces off the surrounding wall N times and encloses an area Θ until it escapes from the billiard table through lead 2. The probability for such an event to occur is the product of the probability p_N that the number of bounces is N and the conditional probability $P_N(\Theta)$ that the area enclosed is Θ for an orbit with fixed N. Both p_N and $P_N(\Theta)$ are determined by the scattering property of a billiard ball showing chaos (see Chapter 3). Since there is no particular constraint on the number of bounces, one should sum up the products $p_N P_N(\Theta)$ over all N, leading to (2.15).

In the above derivation, the shapes of billiard tables are premised to belong to the category in Fig. 1.4(a), where a billiard ball shows a universal feature of full chaos irrespective of individual billiard table shapes. The universal exponential law of the area distribution $P(\Theta)$ in (2.15) is the outcome of the universality of full chaos, and, through (2.14), results in the power spectrum of conductance fluctuations

$$\widehat{C}_D(\eta) \propto (1 + \alpha_{\mathrm{cl}}|\eta|) \exp(-\alpha_{\mathrm{cl}}|\eta|), \tag{2.16}$$

which is also universal for quantum dots with their shapes in the class in Fig. 1.4(a).

It has thus become clear that conductance fluctuations in quantum dots are closely related to billiard-ball dynamics and that the universal feature of chaotic motion is reflected in the universality of conductance fluctuations. One confirms that chaos in classical dynamics should manifest itself in quantum phenomena. The investigation so far has indeed been intuitive, but more rigorous and comprehensive analyses around conductance fluctuations will be developed in Chapters 3 and 4. Readers interested in other important themes on quantum chaos in quantum dots may skip to Chapter 5.

3

MOTION OF A BILLIARD BALL

As indicated in Chapter 2, the application of the semiclassical theory to quantum transport in quantum dots with attached lead wires makes it possible to characterize conductance fluctuations by means of the properties of billiard-ball (point-particle) dynamics. The important property is the distribution of the effectively closed area depicted by a ball coming from the entrance lead and moving between the billiard table walls until leaving through the exit (which may be the same as the entrance). In this chapter, concentrating on a fully chaotic billiard system, both dynamical and statistical treatment of a billiard ball is made to derive a concrete formula for the area distribution. Statistical distributions other than the area distribution will be studied in later chapters. Readers wishing to be rapidly acquainted with the quantum and semiclassical theory of electrical conductance may skip to Chapter 4.

3.1 Expanding Wavefront and Lyapunov Exponent

The billiard ball here means a point particle with no internal degrees of freedom like the spinning motion. We assume that each collision of a particle with the walls is perfectly elastic and is accompanied by no energy dissipation. From this it follows that whenever a particle collides with the walls, it obeys the **law of reflection** that the angle of incidence is equal to that of reflection. Suppose a reference particle starts under a given initial condition (for its position and direction of velocity vector), while its partner starts with the same magnitude of velocity vector under a slightly different initial condition. What will become of the difference of orbits between the pair of particles? Is it possible for two particles with almost identical initial conditions to become separated rapidly as time elapses? To answer the question, one must quantify the rate of rapid separation between a pair of orbits. The description in Sections 3.1 through 3.3 is based mainly on the work by Gaspard[48] and Gaspard and Rice[49].

Let us first consider the case without collisions (see Fig. 3.1). A particle with the initial position q and velocity v and its nearby partner with $q + \delta q$, $v + \delta v$ will show straight motions with their mutual difference growing in time. The difference is proportional to time. If by some events the difference grows much faster than in this case (in general, faster than power laws in time), one may use the terminology of rapid separation or **extreme sensitivity to initial conditions**. Such events occur when particles collide with the billiard table wall.

Consider a curved wall located ahead of the particle motion (see Fig. 3.2). At a collision with the wall, each particle changes its velocity vector abruptly. The relative

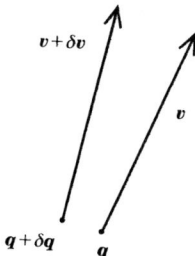

Fig. 3.1   Two nearby orbits in the case of no collision with the wall.

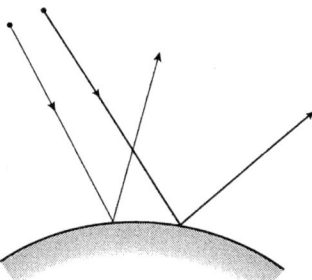

Fig. 3.2   Two nearby orbits in the case of a collision with the wall.

direction and separation between originally nearby orbits also show an abrupt change at their collision with the wall. This change accumulates and the difference of positions and directions between two orbits becomes more and more noticeable as the number of collisions with walls is increased (see Fig. 3.3). Let the separation of positions be magnified by a factor Λ at each collision with the wall. Then the original separation will be multiplied by $\Lambda^n = \exp(n \ln \Lambda)$ after the n-th collision. In other words, the

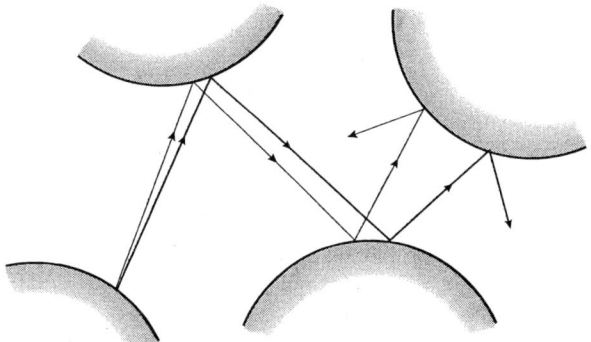

Fig. 3.3   Increase of separation due to successive collisions.

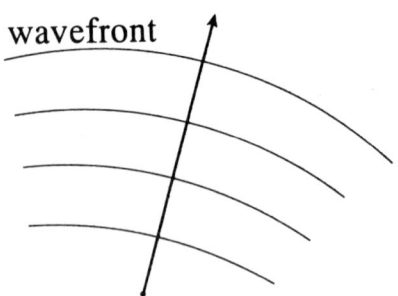

Fig. 3.4   Orbit and curved wavefront.

initially small difference grows **exponentially** by succcessive collisions. This intuitive insight reveals how the presence of curved walls makes a pair of nearby orbits with almost identical initial conditions separate rapidly– a phenomenon of the sensitivity to initial conditions.

It is not true that the multiplication factor is always the same at each collision. This factor, while always larger than unity, depends on the type of collisions or the curvature of the billiard table wall. To see whether or not the separation grows exponentially, therefore, it is necessary to analyze more precisely the nature of each collision and to take a long-time average of the process of such successive scatterings.

Figure 3.4 represents a **wavefront** perpendicular to the particle orbit. An infinitely large number of orbits starting under almost identical conditions (e.g., standing side by side at time $t = 0$) but with the same magnitude of velocity vector form a progressive wavefront. The expanding wavefront conveys the growth of the separation of initially nearby orbits. In what follows, the kinetic energy of a particle is assumed to be kept constant, and therefore the magnitude of the velocity vector is a fixed constant: $|\boldsymbol{v}(t)| = v$ (constant).

For a given reference orbit, consider another orbit with position and velocity increments, $\delta\boldsymbol{q}$ and $\delta\boldsymbol{v}$, almost perpendicular to the reference orbit, i.e., tangential to the wavefront (see also Fig. 3.1).

Before a collision in Fig. 3.1, the velocity increment is conserved,

$$\delta\boldsymbol{v}(t) = \delta\boldsymbol{v}(0),$$

while the position increment increases linearly in time t:

$$\delta\boldsymbol{q}(t) = \delta\boldsymbol{q}(0) + t\delta\boldsymbol{v}(0).$$

The above relations are written in matrix form in terms of $\delta q(t), \delta v(t)$ – components of $\delta\boldsymbol{q}(t), \delta\boldsymbol{v}(t)$ at time t perpendicular to the orbit – as

$$\begin{pmatrix} \delta q(t) \\ \delta v(t) \end{pmatrix} = M_{\text{free}}(t) \begin{pmatrix} \delta q(0) \\ \delta v(0) \end{pmatrix} \tag{3.1}$$

with $M_{\text{free}}(t)$ the matrix for free propagation,

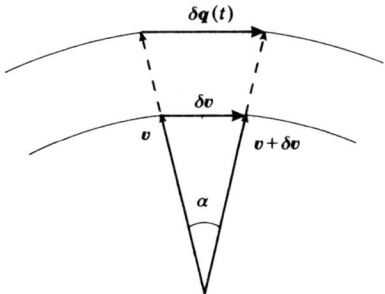

Fig. 3.5 Expanding wavefront during free flight without collision.

$$M_{\text{free}}(t) = \begin{pmatrix} 1 & t \\ 0 & 1 \end{pmatrix}.$$

(3.2)

Figure 3.5 indicates that the difference between two orbits is represented by the angle α formed by them. α itself is determined by the **curvature of the wavefront**. With use of the curvature $\mathcal{B}(t)$ for the wavefront at time t, we find

$$\alpha = \mathcal{B}(t)|\delta\boldsymbol{q}(t)|.$$

(3.3)

Obviously this angle is also given via the velocity increment,

$$\alpha = \frac{|\delta\boldsymbol{v}(t)|}{v}.$$

(3.4)

By eliminating α between (3.3) and (3.4), we obtain an evolution equation for the curvature,

$$\begin{aligned} \mathcal{B}(t) &= \frac{1}{v}\frac{|\delta\boldsymbol{v}(t)|}{|\delta\boldsymbol{q}(t)|} = \frac{1}{v}\frac{|\delta\boldsymbol{v}(0)|}{|\delta\boldsymbol{q}(0) + t\delta\boldsymbol{v}(0)|} \\ &= \frac{1}{vt + \dfrac{v|\delta\boldsymbol{q}(0)|}{|\delta\boldsymbol{v}(0)|}} \\ &= \frac{1}{vt + \dfrac{1}{\mathcal{B}(0)}}, \end{aligned}$$

(3.5)

where $\dfrac{1}{\mathcal{B}(0)}$ stands for the radius of curvature for the initial wavefront. In the above derivation, we exploited the relation

$$|\delta\boldsymbol{q}(0) + t\delta\boldsymbol{v}(0)| = |\delta\boldsymbol{q}(0)| + t|\delta\boldsymbol{v}(0)|$$

which is guaranteed since both $\delta\boldsymbol{q}(t)$ and $\delta\boldsymbol{v}(t)$ are parallel with each other. Equation (3.5) tells us that, during free motions without any collision with walls, the radius of curvature for the wavefront $\left(\dfrac{1}{\mathcal{B}(t)}\right)$ grows at the rate of the velocity v.

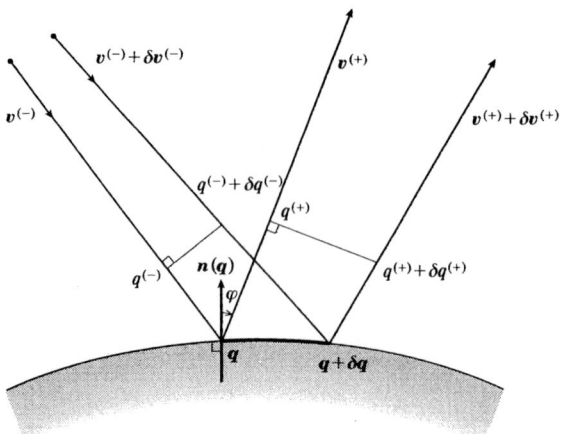

Fig. 3.6   Change of wavefront by collision.

Let us now proceed to a **change of the wavefront caused by collision with a wall**. As illustrated in Fig. 3.6, we investigate the instantaneous change of states just before and after the collision: the change at the walls is assumed to occur abruptly. Let $n(q)$ be an outward normal vector of the wall at the point of collision q, and $v^{(-)}$ and $v^{(+)}$ be the velocity vectors just before and after the collision, respectively. From Fig. 3.6, we easily find the relationship

$$v^{(+)} = v^{(-)} - 2\Big(v^{(-)} \cdot n(q)\Big)n(q) \tag{3.6}$$

with

$$v^{(-)} \cdot n = -v \cos \varphi.$$

For the later discussion, define in general the operator U which only acts on vectors and induces a transformation like (3.6), namely, changing the sign of their component normal to the billiard table boundary and leaving the sign of their tangential component unaffected. In fact, (3.6) can formally be written as

$$v^{(+)} = Uv^{(-)}. \tag{3.7}$$

The meaning of the inverse transformation U^{-1} is self-evident.

Now, we study how the collision event will change the position and velocity increments $\delta q^{(-)}$ and $\delta v^{(-)}$ to $\delta q^{(+)}$ and $\delta v^{(+)}$. The collision does not change the magnitude of $\delta q^{(-)}$. As found from Fig. 3.6, however, the collision at a wall boundary lets a nearby orbit lying to the left side of the reference orbit move to its right side: the relative position vector between the reference and nearby orbits becomes reversed by the collision, giving rise to a sign change in the component of the position increment $\delta q^{(-)}$ perpendicular to the orbit:

$$\delta q^{(+)} = -\delta q^{(-)}. \tag{3.8}$$

On the contrary, the transformation from $\delta v^{(-)}$ to $\delta v^{(+)}$ is a little bit intricate. Let us define another operator V that transforms the position increment $\delta q^{(-)}$ before the collision to that just at the boundary δq (see Fig. 3.6):

$$\delta q = V \delta q^{(-)}. \tag{3.9}$$

Obviously,

$$|\delta q| = \frac{1}{\cos \varphi} |\delta q^{(-)}|.$$

Then we proceed to consider the relation between $v^{(-)} + \delta v^{(-)}$ and $v^{(+)} + \delta v^{(+)}$ available from (3.6):

$$v^{(+)} + \delta v^{(+)} = v^{(-)} + \delta v^{(-)}$$
$$- 2\left\{\left(v^{(-)} + \delta v^{(-)}\right) \cdot n\left(q + \delta q\right)\right\} n\left(q + \delta q\right). \tag{3.10}$$

$n\left(q + \delta q\right)$ implies a modulation of the normal vector caused by a change in the collision point. In the following, we shall expand (3.10) up to terms linear in δq. This expansion can be achieved with use of the **curvature of the billiard table wall** $K(q)$ at the collision point q, and is given by

$$n\left(q + \delta q\right) = n(q) + K(q)\delta q. \tag{3.11}$$

$K(q)$ is positive in the case of the convex wall in Fig. 3.6. In this approximation, the relation (3.10) can be written as

$$v^{(+)} + \delta v^{(+)}$$
$$= v^{(-)} + \delta v^{(-)}$$
$$- 2\left[\left(v^{(-)} + \delta v^{(-)}\right) \cdot \{n(q) + K(q)\delta q\}\right]\{n(q) + K(q)\delta q\}$$
$$= v^{(-)} - 2\left(v^{(-)} \cdot n(q)\right) n(q) + \delta v^{(-)} - 2\left(\delta v^{(-)} \cdot n(q)\right) n(q)$$
$$- 2\left\{\left(v^{(-)} \cdot K(q)\delta q\right) n + \left(v^{(-)} \cdot n(q)\right) K(q)\delta q\right\}. \tag{3.12}$$

Let us recall the transformation operators U and U^{-1}, which act only on vectors and change the sign of their component normal to the billiard table boundary, leaving the sign of their tangential component unaltered. For instance, $U^{-1}n(q) = -n(q)$ and $U^{-1}\delta q = \delta q$. With the help of U and U^{-1}, one can formally rewrite (3.12) as follows:

$$v^{(+)} + \delta v^{(+)}$$
$$= U\left[v^{(-)} + \delta v^{(-)}\right.$$
$$\left. - 2\left\{\left(v^{(-)} \cdot K(q)\delta q\right) U^{-1}n(q) + \left(v^{(-)} \cdot n(q)\right) K(q)U^{-1}\delta q\right\}\right]$$

$$= U \left[v^{(-)} + \delta v^{(-)} \right.$$
$$\left. - 2 \left\{ - \left(v^{(-)} \cdot K(q)\delta q \right) n(q) + \left(v^{(-)} \cdot n(q) \right) K(q)\delta q \right\} \right]$$
$$= U \left[v^{(-)} + \delta v^{(-)} - 2 \left(v^{(-)} \cdot n(q) \right) V^T K(q)\delta q \right] \tag{3.13}$$

where $V^T K(q)\delta q$ is defined by

$$V^T K(q)\delta q \equiv K(q)\delta q - \frac{\left(v^{(-)} \cdot K(q)\delta q \right)}{\left(v^{(-)} \cdot n(q) \right)} n(q). \tag{3.14}$$

One may interpret V^T in (3.14) as a transformation with respect to δq. Since the inner product between $v^{(-)}$ and the right-hand side of (3.14) is vanishing, $V^T K(q)\delta q$ proves perpendicular to the orbit just before collision. This means that V^T is an operator that transforms the vector (δq) along the wall boundary into the one $(v^{(-)})$ perpendicular to the orbit before a collision, indicative of $V^T \propto V^{-1}$. As for the magnitude, noting (see Fig. 3.6)

$$v^{(-)} \cdot \delta q = |v^{(-)}| \, |\delta q| \sin \varphi,$$

we find

$$V^T K(q)\delta q = K(q)|\delta q| \left(\frac{1}{|\delta q|} \delta q + (\tan \varphi) n(q) \right),$$

which yields

$$\left| V^T K(q)\delta q \right| = \frac{1}{\cos \varphi} |K(q)\delta q|.$$

Thanks to the linear nature of U, a combination of (3.7) and (3.13) together with (3.9) and (3.14) leads to the relation between the successive increments,

$$\delta v^{(+)} = U \left[\delta v^{(-)} - 2 \left(v^{(-)} \cdot n(q) \right) V^T K(q) V \delta q^{(-)} \right], \tag{3.15}$$

from which we can derive the transformation from the component $\delta v^{(-)}$ perpendicular to the orbit prior to the collision to the corresponding one $\delta v^{(+)}$ after it. Since the collision merely changes the relative position between a pair of orbits (see Fig. 3.6), the transformation U yields a simple sign change in $\delta v^{(-)}$ and related variables perpendicular to the orbit just before the collision. The role of other transformations, V and V^T, are also obvious. Eventually, so long as the components of increments perpendicular to the orbit are concerned, one may have recourse to the abbreviation

$$U \implies -1, \tag{3.16a}$$

$$V \implies \frac{1}{\cos \varphi}, \tag{3.16b}$$

$$V^T \implies \frac{1}{\cos \varphi}, \tag{3.16c}$$

obtaining the result,

$$\delta v^{(+)} = -\delta v^{(-)} - \frac{2vK(\boldsymbol{q})}{\cos \varphi} \delta q^{(-)}. \tag{3.17}$$

To summarize (3.8) and (3.17), we have

$$\begin{pmatrix} \delta q^{(+)} \\ \delta v^{(+)} \end{pmatrix} = M_{\text{collision}} \begin{pmatrix} \delta q^{(-)} \\ \delta v^{(-)} \end{pmatrix}, \tag{3.18}$$

where the matrix $M_{\text{collision}}$ is defined by

$$M_{\text{collision}} = \begin{pmatrix} -1 & 0 \\ -\dfrac{2vK(\boldsymbol{q})}{\cos \varphi} & -1 \end{pmatrix}. \tag{3.19}$$

With the result (3.18) in mind, we can move on to the investigation of the modulation of the wavefront after successive collisions with billiard table walls. Define the curvature of the wavefront $\mathcal{B}_n^{(+)}$ just after the n-th collision, which, together with the velocity and position increments, $\delta v_n^{(+)}$ and $\delta q_n^{(+)}$, tangential to the wavefront, can provide the angle formed by the reference orbit and its partner (see also (3.3) and (3.4)):

$$\frac{\delta v_n^{(+)}}{v} = \mathcal{B}_n^{(+)} \delta q_n^{(+)}. \tag{3.20}$$

The recursion equation for the successive position and velocity increments tangential to the wavefront just after the n-th and $(n+1)$-th collisions is straightforward from (3.2) and (3.18), and is given by

$$\begin{aligned} \begin{pmatrix} \delta q_{n+1}^{(+)} \\ \delta v_{n+1}^{(+)} \end{pmatrix} &= M_{\text{collision}} M_{\text{free}}(t_n) \begin{pmatrix} \delta q_n^{(+)} \\ \delta v_n^{(+)} \end{pmatrix} \\ &= \begin{pmatrix} -1 & -t_n \\ -\dfrac{2vK_n}{\cos \varphi_n} & -1 - 2\dfrac{vK_n t_n}{\cos \varphi_n} \end{pmatrix} \begin{pmatrix} \delta q_n^{(+)} \\ \delta v_n^{(+)} \end{pmatrix}. \end{aligned} \tag{3.21}$$

Here, M_{free} was obtained in (3.2), K_n and φ_n denote the curvature of the wall and angle of reflection, respectively, at the n-th collision point, and t_n is the time for a free flight during the n-th and $(n+1)$-th collisions. (One should not be confused between the curvature of the wall K_n and that of the wavefront $\mathcal{B}_n^{(+)}$.)

Using (3.20) in (3.21), one has

$$
\begin{pmatrix} \delta q_{n+1}^{(+)} \\ v\mathcal{B}_{n+1}^{(+)} \delta q_{n+1}^{(+)} \end{pmatrix} = \begin{pmatrix} -1 & -t_n \\ -\dfrac{2vK_n}{\cos\varphi_n} & -1 - 2\dfrac{vK_n t_n}{\cos\varphi_n} \end{pmatrix} \begin{pmatrix} \delta q_n^{(+)} \\ v\mathcal{B}_n^{(+)} \delta q_n^{(+)} \end{pmatrix}.
$$

$$(3.22)$$

Upper and lower equalities obtained from the above matrix equation are not independent of each other, and, eliminating $\delta q_n^{(+)}$ and $\delta q_{n+1}^{(+)}$ between these two equalities, we get an evolution equation for the curvature of the wavefront,

$$
\begin{aligned}
v\mathcal{B}_{n+1}^{(+)} &= \frac{-\dfrac{2vK_n}{\cos\varphi_n} + \left(-1 - 2\dfrac{vK_n t_n}{\cos\varphi_n}\right) v\mathcal{B}_n^{(+)}}{-1 - t_n v\mathcal{B}_n^{(+)}} \\
&= \frac{2vK_n}{\cos\varphi_n} + \frac{1}{t_n + \dfrac{1}{v\mathcal{B}_n^{(+)}}}.
\end{aligned}
$$

$$(3.23)$$

On the basis of this formula, one can calculate the curvature of the wavefront that is perpendicular to orbits each time a particle bounces off the wall boundary. One can now quantify the degree of spreading of the wavefront because its radius of curvature is an inverse of the curvature. And ultimately, one comes to capture exactly how a pair of orbits starting under nearly identical initial conditions will get separated after infinite repetitions of collision with billiard table walls.

Let us now embark upon defining the **Lyapunov exponent** in terms of the curvature. This exponent characterizes the averaged rate at which the distance between orbits between a pair of initially adjacent orbits grows as time elapses. Consider two orbits which have almost identical positions and velocity vectors. Their kinetic energies and therefore the magnitudes of the velocity vector are assumed to be the same, which ensures that the difference between velocity vectors is perpendicular to the orbits. Let T_n be the time that orbits bounce at the n-th wall and t_{n-1} be the time interval for a free flight during the $(n-1)$-th and n-th collisions. The distance $|\delta q(T_n)|$ between orbits at time T_n is

$$
\begin{aligned}
|\delta q(T_n)| &= |\delta q(T_{n-1} + t_{n-1})| \\
&= |\delta q(T_{n-1}) + t_{n-1} \delta v_{n-1}|.
\end{aligned}
$$

$$(3.24)$$

Substituting (3.20) into (3.24), we have

$$
\begin{aligned}
|\delta q(T_n)| &= \left|\left(1 + vt_{n-1}\mathcal{B}_{n-1}^{(+)}\right)\right| |\delta q(T_{n-1})| \\
&= \left|\left(1 + vt_{n-1}\mathcal{B}_{n-1}^{(+)}\right)\right| \left|\left(1 + vt_{n-2}\mathcal{B}_{n-2}^{(+)}\right)\right| \\
&\quad \cdots \left|\left(1 + vt_0\mathcal{B}_0^{(+)}\right)\right| |\delta q(0)|,
\end{aligned}
$$

$$(3.25)$$

which shows how the initial small deviation gets magnified by repeated collisions with the walls.

The definition of the **Lyapunov exponent** λ is provided by assuming an exponential growth of the initial increment in the long time limit as

$$|\delta q(t)| \sim \exp(\lambda t)|\delta q(0)|.$$

Comparing this definition with (3.25), we reach the formula

$$\lambda = \lim_{N \to \infty} \left(\frac{\sum_{n=1}^{N} \ln \left|1 + t_n v \mathcal{B}_n^{(+)}\right|}{\sum_{n=1}^{N} t_n} \right). \tag{3.26}$$

$v\mathcal{B}_n^{(+)}$ is calculable from (3.23) and the initial curvature may be chosen as

$$v\mathcal{B}_1^{(+)} = \frac{2vK_0}{\cos \varphi_0} + \frac{1}{t_0}, \tag{3.27}$$

by assuming initially a negligibly small position increment. In the case that the curvature K of the billiard table walls is always positive,

$$\ln \left|1 + t_n v \mathcal{B}_n^{(+)}\right| > 0$$

is guaranteed for all n, which immediately gives a positive Lyapunov exponent. In this way, the analysis of the expanding wavefront elucidates how a pair of nearby orbits gets separated in due course.

3.2 Birkhoff Coordinates and the Repeller

In the previous section, we treated the situation where a particle is completely confined to billiard systems – closed systems. We there assumed billiard systems such as a stadium or the Sinai billiard consisting of an infinite number of convex hard disks so that the particle can repeat bouncing with the walls an infinite number of times. In this section, on the contrary, we shall pay attention to open billiard systems accompanied by windows or holes through which a particle can escape sooner or later.

To make an open billiard system imaginable, we choose a system of three disk scatterers in Fig. 3.7. These disks are placed so as to make openings (space) through which a particle comes in or gets out. What is the fate of a particle launched at some position outside towards the scatterers? A pair of orbits with almost identical initial conditions gets separated rapidly by repetition of their bouncing with the walls. If the system has leaky openings, however, a particle may get out through them or stay for a long time inside the system.

In billiard systems, each orbit is determined geometrically, and the kinetic energy or the magnitude of the velocity vector ($v = |\boldsymbol{v}|$) of a particle have nothing to do with determining the orbits. So, in the following we prescribe $v = 1$ and bring in convenient coordinates. Choose some point on the billiard table boundary as the origin and let the

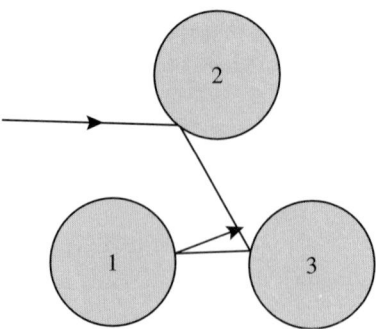

Fig. 3.7   Scattering of a particle by three disk scatterers.

length s_n along the billiard table wall (in a clockwise direction) from the origin denote the n-th bouncing point of a particle (see Fig. 3.8). At this point we also introduce the angle φ_n between the direction of incidence and that of the inwardly normal vector of the curved wall, which is nothing but the angle of incidence. A sequence of values for a pair of coordinates s_n and $\sin\varphi_n$ completely captures the billiard-ball dynamics in phase space. Further, the map for successive values for s_n and $\sin\varphi_n$ preserves phase-space area, and is convenient for analysis of the distribution of orbits in phase space. This pair of coordinates is called the **Birkhoff pair of coordinates** and the map on Birkhoff coordinates as the **Birkhoff map**.

Here we shall give a proof of the area-preserving property of the Birkhoff map. Let δs and $\delta\varphi$ be increments of the collision point on the wall and that of the angle of incidence, respectively. A sequence of their values is generated through succesive collisions, and we can derive a linear map for them with the help of components of position and velocity increments perpendicular to the orbit.

Figure 3.9 shows that the perpendicular component $\delta q_0^{(-)}$ of a position increment prior to collision causes an increment of the collision point by

$$\delta s_0 = \frac{1}{\cos\varphi_0}\delta q_0^{(-)}.$$

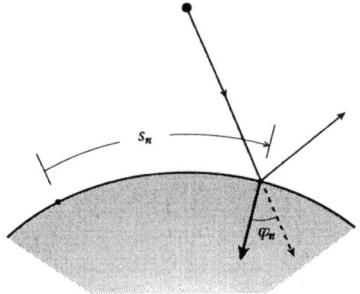

Fig. 3.8   Birkhoff coordinates.

As for the increment $\delta\varphi_0$ of the angle of incidence, it will be generated by perpendicular components $\delta q_0^{(-)}$ and $\delta v_0^{(-)}$. Noting

$$\varphi_0 + \delta\varphi_0 = \varphi_0 + \delta v_0^{(-)} + K(s_0)\delta s_0,$$

we obtain

$$\delta\varphi_0 = K(s_0)\delta s_0 + \delta v_0^{(-)}.$$

Both results above are summarized in matrix form as the transformation from $\delta s_0, \delta\varphi_0$ to $\delta q_0^{(-)}, \delta v_0^{(-)}$:

$$\begin{pmatrix} \delta q_0^{(-)} \\ \delta v_0^{(-)} \end{pmatrix} = \begin{pmatrix} \cos\varphi_0 & 0 \\ -K(s_0) & 1 \end{pmatrix} \begin{pmatrix} \delta s_0 \\ \delta\varphi_0 \end{pmatrix}. \tag{3.28}$$

In a similar way, Fig. 3.10 shows how the perpendicular components $\delta q_1^{(+)}, \delta v_1^{(+)}$ of position and velocity increments just after a collision are related to the increments $\delta s_1, \delta\varphi_1$ at the collision point, giving the relation

$$\begin{pmatrix} \delta s_1 \\ \delta\varphi_1 \end{pmatrix} = \begin{pmatrix} -\dfrac{1}{\cos\varphi_1} & 0 \\ \dfrac{K(s_1)}{\cos\varphi_1} & -1 \end{pmatrix} \begin{pmatrix} \delta q_1^{(+)} \\ \delta v_1^{(+)} \end{pmatrix}. \tag{3.29}$$

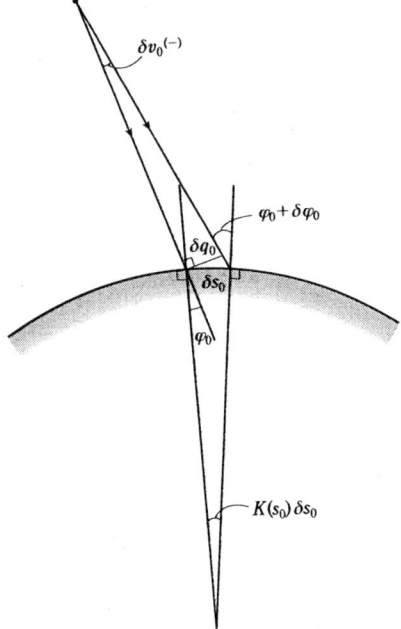

Fig. 3.9   Incident orbits and curved wall.

A sequence of transformations, $(3.28) \to M_{\text{collision}} \to M_{\text{free}}(t_{01}) \to M_{\text{collision}} \to$ (3.29), with t_{01} the time for free flight between the first and second collisions establishes a linear map from $\delta s_0, \delta \varphi_0$ to $\delta s_1, \delta \varphi_1$:

$$
\begin{pmatrix} \delta s_1 \\ \delta \varphi_1 \end{pmatrix} = \begin{pmatrix} -\dfrac{1}{\cos \varphi_1} & 0 \\ \dfrac{K(s_1)}{\cos \varphi_1} & -1 \end{pmatrix} \begin{pmatrix} -1 & 0 \\ -\dfrac{2K(s_1)}{\cos \varphi_1} & -1 \end{pmatrix} \begin{pmatrix} 1 & t_{01} \\ 0 & 1 \end{pmatrix}
$$

$$
\times \begin{pmatrix} -1 & 0 \\ -\dfrac{2K(s_0)}{\cos \varphi_0} & -1 \end{pmatrix} \begin{pmatrix} \cos \varphi_0 & 0 \\ -K(s_0) & 1 \end{pmatrix} \begin{pmatrix} \delta s_0 \\ \delta \varphi_0 \end{pmatrix}
$$

$$
= -\begin{pmatrix} \dfrac{1}{\cos \varphi_1} & 0 \\ \dfrac{K(s_1)}{\cos \varphi_1} & 1 \end{pmatrix} \begin{pmatrix} 1 & t_{01} \\ 0 & 1 \end{pmatrix} \begin{pmatrix} \cos \varphi_0 & 0 \\ K(s_0) & 1 \end{pmatrix} \begin{pmatrix} \delta s_0 \\ \delta \varphi_0 \end{pmatrix},
$$

$$(3.30)$$

which can be reduced to the linealized Birkhoff map $(M_{1 \leftarrow 0})$ acting on increments $(\delta s, \delta \sin \varphi)$ of Birkhoff coordinates,

$$
\begin{pmatrix} \delta s_1 \\ \delta(\sin \varphi_1) \end{pmatrix} = M_{1 \leftarrow 0} \begin{pmatrix} \delta s_0 \\ \delta(\sin \varphi_0) \end{pmatrix}
$$

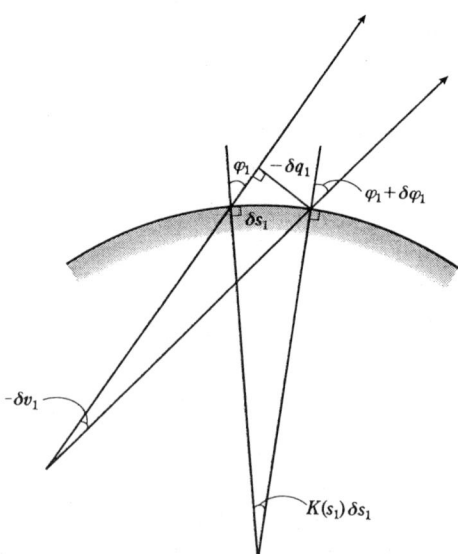

Fig. 3.10   Outgoing (reflecting) orbits and curved wall.

$$\equiv -\begin{pmatrix} 1 & 0 \\ 0 & \cos\varphi_1 \end{pmatrix} \begin{pmatrix} \dfrac{1}{\cos\varphi_1} & 0 \\ \dfrac{K(s_1)}{\cos\varphi_1} & 1 \end{pmatrix} \begin{pmatrix} 1 & t_{01} \\ 0 & 1 \end{pmatrix}$$

$$\times \begin{pmatrix} \cos\varphi_0 & 0 \\ K(s_0) & 1 \end{pmatrix} \begin{pmatrix} 1 & 0 \\ 0 & \dfrac{1}{\cos\varphi_0} \end{pmatrix} \begin{pmatrix} \delta s_0 \\ \delta(\sin\varphi_0) \end{pmatrix}.$$

$$(3.31)$$

Obviously, $\det(M_{1\leftarrow 0}) = -1$, ensuring that the Birkhoff map is **area-preserving**. As a consequence, if the particle dynamics is completely chaotic, the phase space spanned by Birkhoff coordinates is fully occupied with a chaotic sea.

Figure 3.11 displays the phase-space structure for particle dynamics in the three disk scatterers (unit disks) in Fig. 3.7. Figure 3.11(a) shows states of departure for an assembly of orbits (black belts) that are destined to hit the next scatterers. Figure 3.11(b) indicates how the assembly of initial points in Fig. 3.11(a) is transformed by the first collision. For instance, the black belt • 12 in Fig. 3.11(a) is moved to one 1 • 2 in Fig. 3.11(b). In each black belt, numbers 1, 2, 3 denote the disks in Fig. 3.7, and the number just behind "•" indicates the disk number at which the collision is currently occurring. With this prescription, each orbit can be symbolically represented by a sequence of numbers. The first collision drives the black belts, • 12 and • 13, in Fig. 3.11(a) to 1 • 2 and 1 • 3 in Fig. 3.11(b), respectively. Namely, orbits bounced at the disk 1 are decomposed into three families, one hitting disk 2, another hitting disk 3, and the other escaping out of the scatterers.

What is the future of orbits 1 • 2 and 1 • 3 in Fig. 3.11(b) after the second collision? Noting that only the black belts in disk 2 can meet the next scatterer, a fraction of the black belt 1 • 2 that overlaps with • 21 and • 23 in Fig. 3.11(a) can continue to dwell inside the scatterers. After repeating this procedure an infinite number of times, most orbits escape outside, leaving the measure-zero orbits called **repellers** trapped for a long time. The repellers constitute the Cantor set leading to a fractal structure. Thus, the particle dynamics in open billiards leaves repellers that will be responsible for both classical and quantum transport.

3.3 Kolmogorov–Sinai Entropy and Escape Rate

The next strategy is to quantify the way a particle escapes outside the billiard system through openings (windows) and to characterize the stochastic property of the motion of repellers. **Kolmogorov–Sinai (KS) entropy** is a convenient measure of deterministic randomness, which is defined through **algorithmic complexity**. Consider a given time sequence ω_t with time interval t. Can we make a conjecture on the succeeding time sequences? In the case that the original sequence ω_t is periodic, one can predict the future sequences perfectly with the use of the one-period data in ω_t. On the other hand, if ω_t is a random sequence, one must continue to record the

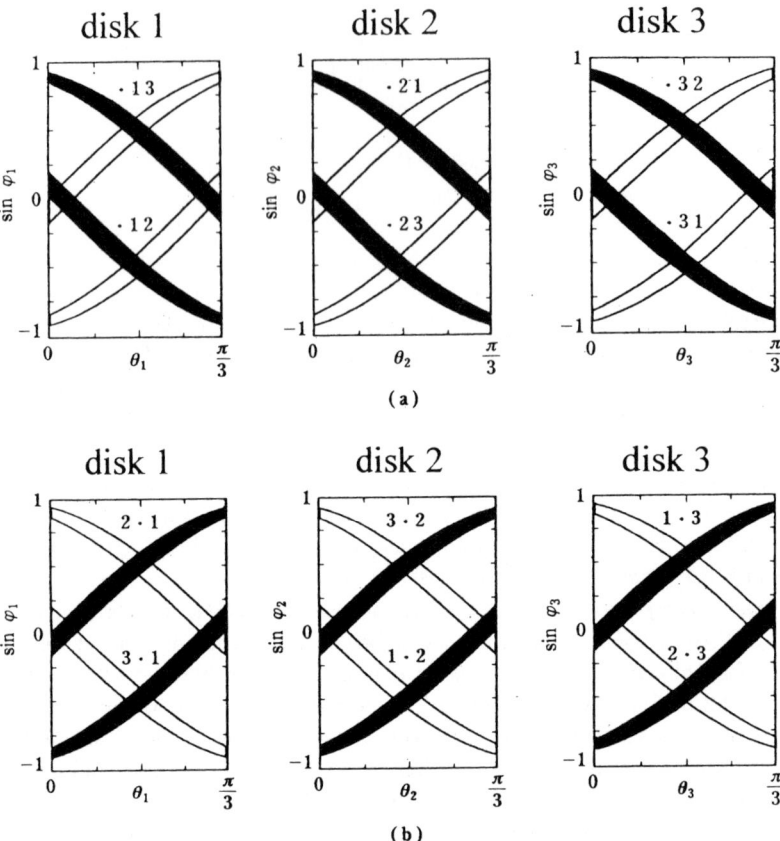

Fig. 3.11   Phase-space structure for particle dynamics in the three disk scatterers (disks) in Fig. 3.7. A set of three horizontal figures completes the phase space with Birkhoff coordinates defined for each scatterer. The radius of each disk is assumed to be unity. On abscissa, angles at the circumference $\theta_j (j = 1, 2, 3)$ stand for lengths $s_j (j = 1, 2, 3)$ along the circular arc: (a) an assembly of initial points (black belts) that inevitably hit the next scatterers. Each black belt has its time-reversal partner (white belts); (b) new black belts as the image of the transformation (via Birkhoff mapping) of the original ones in (a) (reproduced from[48]).

succeeding data.

Let us imagine writing a computer program to reproduce the data in ω_t. In the case of the time-periodic sequence, the algorithm is simple and short. However, in the case of the random sequence, there is no way other than writing the data themselves, which is lengthy. Thus, the length of an algorithm is a measure of randomness inherent to the time sequence. The **algorithmic complexity** $K(\omega_t)$ for a given time sequence ω_t is defined as the length of the shortest program to describe ω_t. For a completely random time sequence, the program to generate it is just writing this sequence; the length of the program is proportional to t, i.e., the length of ω_t. The randomness of ω_t will be quantified in terms of the KS entropy: with the knowledge on how the algorithmetic complexity grows with increasing the length of the time sequence, the KS entropy h_{KS} is defined as

$$h_{\mathrm{KS}} = \lim_{t \to \infty} \frac{K(\omega_t)}{t}. \tag{3.32}$$

For the completely random time sequence, h_{KS} takes a finite positive value because the algorithmic complexity grows linearly in time.

Coming back to billiard-ball dynamics, randomness of the time sequence generated by a deterministic law is also characterized by KS entropy. For the chaotic motion of a billiard ball, the **sensitivity to initial conditions** leads to the exponential growth of the initially small error (increment). A succcession of abrupt expansions of the wavefront occurring at its collision with wall boundaries causes a rapid separation of nearby orbits. This means that, even under the extremely precise preparation of initial conditions, the position of a billiard ball after long time is unpredictable and stochastic, generating a random time sequence. **Partitioning of the phase space** can reveal the stochasticity involved in the initial condition and makes it possible to attribute KS entropy to billiard ball dynamics.

Let the phase space spanned by Birkhoff coordinates be partitioned into n cells. Each cell includes a lot of points that represent the position and velocity vector of a billiard ball. Once individual points in a given cell could be distinguished from each other, the cells to which these points transmit will be determined without any stochastic mechanism. Practically, however, our measurement has a limited resolution, and we can only identify each cell where the phase-space point belongs, and the transmission to another cell is probabilistic.

How is the above situation altered with increasing the number of partitions or with decreasing the cell size? In the case of regular dynamics where nearby orbits show no rapid separation, all points in a given cell can transmit as a whole to another cell so long as small-enough cells are employed, so each cell-to-cell transmission is not probabilistic. In contrast, in the case of chaotic dynamics where nearby orbits show rapid separation, points in a given cell can transmit to different cells even when small-enough cells are employed: the cell-to-cell transmission is probabilistic against any choice of small-enough cells.

In this way, partitioning of the phase space into cells makes billiard ball dynamics to be represented symbolically as a time sequence, and thereby to be characterized in terms of KS entropy. A detailed insight yields a more practical definition of KS entropy than (3.32) as follows:

$$h_{\mathrm{KS}} = \lim_{n \to \infty} \left(-\frac{\sum_{a,b} \pi^{(n)}(a) P_{ab}^{(n)} \ln P_{ab}^{(n)}}{\sum_{a} t_{ab}^{(n)} \pi^{(n)}(a)} \right), \tag{3.33}$$

where n is the number of whole cells or the number of partitioning the phase space, and a and b denote individual cells; $P_{ab}^{(n)}$ is the probability for a point in one cell a

to transmit to another b via the Birkhoff map; $\pi^{(n)}(a)$ denotes the probability of the repellers to be in the cell a; $t_{ab}^{(n)}$, common to all terminal cells b, is the free flight time of a particle responsible for the cell-to-cell transport. If, for finer and finer partitions, the cell-to-cell transition probability will become $P_{ab}^{(n)} \to 1$ or $P_{ab}^{(n)} \to 0$, the definition (3.33) leads to $h_{KS} = 0$. On the other hand, if the limiting behavior of $P_{ab}^{(n)}$ will show a broad spectrum, one can expect $h_{KS} > 0$.

In open billiards under consideration, we must introduce another important quantity. Here a particle comes in or leaves through the holes or windows. Let an assembly of particles be initially distributed inside a billiard system. Particles begin to move in various directions, and, under the assumption of no interaction among them, either escape or stay as the repellers inside the systems. In the long-time limit, the number of particles remaining inside the billiard system decays exponentially in time with the decay rate called the **escape rate**. The escape rate quantifies a feature of the phase-space points near the repellers to go away from them. In what follows we shall explain how the escape rate is to be related with the Lyapunov exponent – a measure of the orbital instability – and KS entropy – a measure to quantify the stochastic aspect of orbits.

Consider an ensemble of points gathered around a cell a in phase space. This distribution diffuses by repeated collisions. Let us assume that the n-th collision partially transforms the distribution $p_{a,n-1}$ at the cell a to the distribution $q_{b,n}$ at the cell b. The transformation between these distributions satisfies

$$q_{b,n} = p_{a,n-1} Q_{ab}^{(n)}.$$

Collecting the trasformations to the cell b from all cells in the phase space, one has the distribution $p_{b,n}$ at the cell b,

$$p_{b,n} = \sum_a q_{b,n} = \sum_a p_{a,n-1} Q_{ab}^{(n)}.$$

In the above, $Q_{a,b}^{(n)}$ is the transition rate governed by the stretching rate, Λ_{ab}, for the distribution around the cell a:

$$Q_{ab}^{(n)} = \frac{1}{\Lambda_{ab}}. \tag{3.34}$$

The rate Λ_{ab} is determined by the Birkhoff map, whether billiards are open or closed: the larger the rate Λ_{ab}, the smaller the transition rate from cell a to b.

In the open systems, the definition of $Q_{ab}^{(n)}$ is more intricate. In a long-time regime, the phase-space points consist of repellers and their neighbors that leave the phase space sooner or later. The repellers, which have zero measure in the whole phase space and constitute a Cantor set, have a finite probability $\pi^{(n)}(a)$ invariant against particle collisions with walls (see Fig. 3.12). For the repellers, the cell-to-cell transition rate $Q_{ab}^{(n)}$ should be magnified by a multiplication factor that compensates the loss of distributions due to escaping through openings. With the use of the escape rate γ

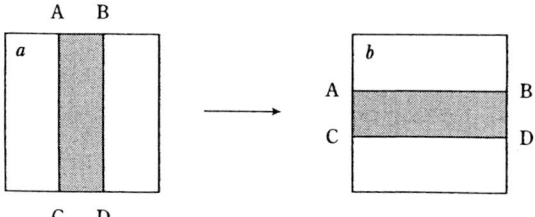

Fig. 3.12 Transition from one cell a to another b. Billiard ball dynamics brings the gray area in a to that in b.

and the cell-to-cell flight time t_{ab}, the improved transition rate for repellers is defined as

$$Q_{ab}^{(n)} = \frac{\exp(\gamma t_{ab})}{\Lambda_{ab}}.$$ (3.35)

Since the probability distribution on the repellers is invariant against each collision, a conservation law should hold for each of the time evolutions for the future and the past:

$$\sum_a u_a^{(n)} Q_{ab}^{(n)} = u_b^{(n)},$$ (3.36a)

$$\sum_b Q_{ab}^{(n)} v_b^{(n)} = v_a^{(n)},$$ (3.36b)

where $u_a^{(n)}$ and $v_a^{(n)}$ are distributions proper to forward and backward time evolutions, respectively, and the invariant probability $\pi^{(n)}(a)$ is defined by their product,

$$\pi^{(n)}(a) = u_a^{(n)} v_a^{(n)}.$$

To evaluate the KS entropy in (3.33), the cell-to-cell **transition probability** for the repellers (see Fig. 3.12) is needed, which is now given as

$$P_{ab}^{(n)} = \frac{Q_{ab}^{(n)} v_b^{(n)}}{v_a^{(n)}}.$$ (3.37)

Combining (3.33) with (3.37) and the equation below (3.36b), we have

$$h_{\mathrm{KS}} = -\frac{1}{\sum_a t(a)\pi(a)} \sum_{a,b} u_a v_a \frac{Q_{ab} v_b}{v_a} \ln\left(\frac{Q_{ab} v_b}{v_a}\right)$$

$$= -\frac{1}{\sum_a t(a)\pi(a)} \sum_{a,b} u_a Q_{ab} v_b \left[\gamma t(a) - \ln \Lambda_a + \ln v_b - \ln v_a\right],$$

(3.38)

where, noting that t_{ab} are common to all terminal cells b, the prescription $t(a) = t_{ab}$ is employed. With help of the following relations available from (3.36a) and (3.36b),

$$\sum_{a,b} u_a Q_{ab} v_b \gamma t(a) = \sum_a \gamma \pi(a) t(a),$$ (3.39a)

$$\sum_{a,b} u_a Q_{ab} v_b \ln v_a = \sum_a \pi(a) \ln v_a, \tag{3.39b}$$

$$\sum_{a,b} u_a Q_{ab} v_b \ln v_b = \sum_b \pi(b) \ln v_b, \tag{3.39c}$$

(3.38) becomes

$$h_{\mathrm{KS}} = -\gamma + \frac{\sum_a \pi(a) \ln \Lambda_a}{\sum_a t(a)\pi(a)}. \tag{3.40}$$

In the second term on the right-hand side of (3.40), the denominator represents the phase space average of the free flight time between succcesive collisions, and the numerator is the corresponding average of the local Lyapunov exponent. Namely, this second term is nothing but the Lyapunov exponent λ per unit time. Hence, the escape rate, Lyapunov exponent and KS entropy obey the equality (Kantz and Grassberger[50], Eckmann and Ruelle[51], Gaspard[48]),

$$\gamma = \lambda - h_{\mathrm{KS}}. \tag{3.41}$$

In the specific limit of closed billiards with $\gamma = 0$, (3.41) reduces to Pesin's formula,

$$h_{\mathrm{KS}} = \lambda. \tag{3.42}$$

The positive Lyapunov exponent implies a rapid separation of nearby orbits as time elapses. Then the small uncertainty – uncontrollable increments – in the initial position and velocity vector gives rise to the probabilistic feature of the billiard ball dynamics with positive KS entropy. This is what Pesin's formula means.

In the case of open billiards, most of the orbits get away through openings due to their ergodicity. In the long-time limit, only the repellers (e.g., periodic orbits and asymptotic orbits continuing to approach them) survive, which have a positive Lyapunov exponent. As suggested from the apparently diminished KS entropy in (3.41) due to the positive escape rate, however, the probabilistic feature of repellers is more or less suppressed. The positive escape rate thus plays an important role in characterizing open billiards.

3.4 Area Distribution

As we saw in the above, the escape rate is an important concept inherent to repellers in an open billiard system – a billiard system with openings. On the other hand, the semiclassical theory relates the **area distribution** to the conductance in quantum billiards with attached lead wires as openings. This section is devoted to a derivation of the relationship between the escape rate and the area distribution (see Fig. 3.13). The **nature of the repellers will show up in the conductance fluctuations**.

For an open billiard system, let the number $N(t)$ of particles dwelling inside decrease exponentially in time with the escape rate γ, namely,

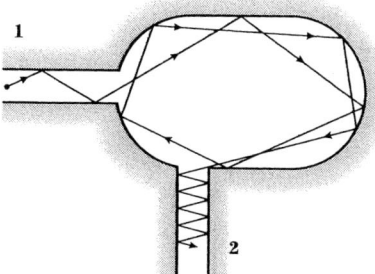

Fig. 3.13 Transmitting orbit from lead 1 to lead 2.

$$\frac{N(t)}{N(0)} \sim \exp(-\gamma t). \tag{3.43}$$

The flight time gives the flight distance $l = vt$ with v fixed to some common constant value for a given Fermi energy E_{F}. The number $\widetilde{N}(l)$ of particles that stay inside the billiard system after traveling over the distance l is straightforward from (3.43):

$$\frac{\widetilde{N}(l)}{N(0)} \sim \exp\left(-\frac{\gamma}{v}l\right). \tag{3.44}$$

Let $\bar{l}$ be the mean distance between successive collisions. Then the number of collisions for a given flight distance l is $N = \dfrac{l}{\bar{l}}$ and (3.44) gives the distribution for N:

$$P(N) \propto \exp\left(-\widetilde{\gamma}N\right) \tag{3.45}$$

with $\widetilde{\gamma} = \frac{\gamma}{v}\bar{l}$. (3.45) in turn gives the distribution for the number N of particles that escape outside after colliding with the walls N times,

$$\widetilde{P}(N) = P(N) - P(N+1)$$
$$\propto (1 - \exp(-\widetilde{\gamma}))\exp(-\widetilde{\gamma}N). \tag{3.46}$$

Finally the probability p_N that an incident particle hits the walls N times before escaping out of the cavity is proportional to $\widetilde{P}(N)$, and we arrive at

$$p_N \propto \exp(-\widetilde{\gamma}N). \tag{3.47}$$

To obtain the **effective area distribution** with use of the probability p_N, we need a conditional probability $P_N(\Theta)$ for the area Θ under a fixed N. Taking the three disk scatterers in Fig. 3.7 as a prototype for an open chaotic billiard system, we shall proceed with the investigation. In Fig. 3.14, let O be the origin of the system and consider an example of successive bounces of a particle from one point X to another point Y. This flight has a tendency to show a clockwise rotation, since the position vector $\overrightarrow{OX}$ is changed to $\overrightarrow{OY}$. In general, each orbit has an effective direction

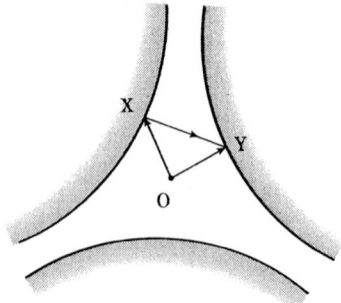

Fig. 3.14   An orbit and its direction of rotation.

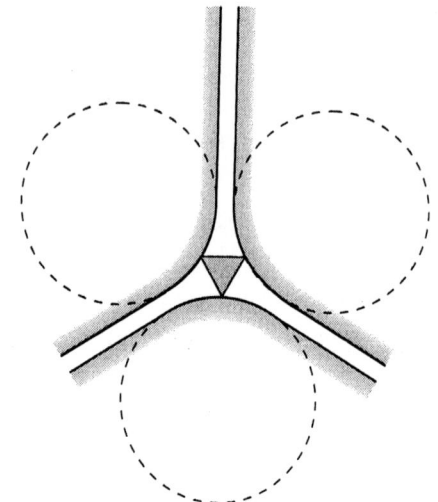

Fig. 3.15   Collision number, length and area for transmission orbit.

of rotation. For an orbit hitting the walls N times until leaving the system, let N_{R} and N_{L} be the numbers of clockwise and counter-clockwise rotations, respectively. Noting the mean distance $\bar{l}$ of a flight during successive collisions, the topological length executed for a particle before going away from the exit is given by

$$l_{topol} = (N_{\mathrm{R}} - N_{\mathrm{L}})\bar{l} \equiv m\bar{l} \tag{3.48}$$

with m the net rotation number.

Each billiard system has its proper collision number $\overline{N}$: an orbit colliding with the walls $\overline{N}$ times forms a proper polygon with $\overline{N}$ vertices and mean area $\overline{\Theta}$. In the example in Fig. 3.15, the polygon is a triangle with $\overline{N} = 3$. Such an orbit runs over the distance $\overline{N} \times \bar{l}$. In general, an orbit with its topological length l_{topol} encloses this mean area $\dfrac{l_{topol}}{\overline{N}\,\bar{l}}$ times. Eventually, an orbit with collision number N and topological length l_{topol} encloses the effective area Θ given by

$$\Theta = \overline{\Theta}\frac{l_{topol}}{\overline{N}\,\overline{l}} = \frac{m}{\overline{N}}\overline{\Theta}. \tag{3.49}$$

Θ is positive for clockwise rotation. From (3.49), we see that once the distribution for the net rotation number $m(\equiv N_{\mathrm{R}} - N_{\mathrm{L}})$ with a fixed N is available, the area distribution $P_N(\Theta)$ with a fixed N will also be obtained. Consider a Gaussian probability distribution $\mathcal{P}_N(m)$ for m with mean zero and variance N:

$$\mathcal{P}_N(m) = \frac{1}{\sqrt{2\pi N}}\exp\left(-\frac{m^2}{2N}\right). \tag{3.50}$$

With help of the variable transformation (3.49) together with its derivative $dm = \frac{\overline{N}}{\overline{\Theta}}d\Theta$, $\mathcal{P}_N(m)$ can be transformed to the area distribution $P_N(\Theta)$ with a fixed N as

$$
\begin{aligned}
P_N(\Theta) &= \mathcal{P}_N(m)\frac{dm}{d\Theta} \\
&= \frac{1}{\sqrt{2\pi N}}\frac{\overline{N}}{\overline{\Theta}}\exp\left(-\frac{m^2}{2N}\right) \\
&= \frac{1}{\sqrt{2\pi\left(\dfrac{N}{\overline{N}^2}\overline{\Theta}^2\right)}}\exp\left(-\frac{\Theta^2}{2\left(\dfrac{N}{\overline{N}^2}\overline{\Theta}^2\right)}\right).
\end{aligned} \tag{3.51}
$$

The probability distribution $P(\Theta)$ for the effective area Θ is given as a summation of the product of p_N in (3.47) and $P_N(\Theta)$ in (3.51) over all N. Replacing the sum by an integration, we have

$$
\begin{aligned}
P(\Theta) &\sim \sum_{N=1}^{\infty} P_N(\Theta)p_N \\
&\sim \int_1^{\infty} \frac{1}{\sqrt{2\pi\left(\dfrac{N}{\overline{N}^2}\overline{\Theta}^2\right)}}\exp\left(-\frac{\Theta^2}{2\left(\dfrac{N}{\overline{N}^2}\overline{\Theta}^2\right)} - \widetilde{\gamma}N\right)dN.
\end{aligned} \tag{3.52}
$$

The above integral can be evaluated in the saddle point approximation: Noting the saddle point

$$N^* = \frac{|\Theta|}{\sqrt{2\widetilde{\gamma}}\,\overline{\Theta}}\overline{N}$$

and introducing the deviation $x \equiv N - N^*$, we reach the goal (aside from a normalization factor):

$$P(\Theta) \sim \frac{1}{\sqrt{2\pi\left(\dfrac{N^*}{\overline{N}^2}\overline{\Theta}^2\right)}}\exp\left(-\frac{\Theta^2}{2\left(\dfrac{N^*}{\overline{N}^2}\overline{\Theta}^2\right)} - \widetilde{\gamma}N^*\right)$$

$$\times \int_{-\infty}^{\infty} \exp\left(-\frac{1}{2}\frac{\overline{\Theta}}{\overline{N}\Theta}\left(\sqrt{2\widetilde{\gamma}}\right)^3 x^2\right) dx$$

$$= \frac{\overline{N}}{\sqrt{2\widetilde{\gamma}\,\overline{\Theta}}} \exp\left(-\frac{\overline{N}}{\overline{\Theta}}\sqrt{2\widetilde{\gamma}}\,|\Theta|\right). \tag{3.53}$$

This exponential area distribution has been derived only by assuming that the billiard ball dynamics is chaotic. While system-specific quantities $\overline{\Theta}$, $\overline{N}$, $\widetilde{\gamma}$ are included in (3.53), the exponential functional form is universal and proper to chaotic billiards listed in Fig. 1.4(a) with openings. This functional form originates in the fact that the number of particles trapped inside the billiard system decreases exponentially in time. In other words, in the case of chaotic billiards in Fig. 1.4(a) with openings, both the Lyapunov exponent and the KS entropy have positive values; the exponential area distribution is then inevitable to guarantee the nonzeo escape rate ruled by the formula (3.41). This universal area distribution proper to open chaotic billiards will show up as a universal conductance fluctuations in quantum transport. This point will be elucidated in the next chapter.

4

SEMICLASSICAL THEORY OF CONDUCTANCE FLUCTUATIONS

In this chapter, after the explanation of quantum theory on conductance fluctuations in quantum dots, we shall illustrate a semiclassical method to describe the conductance as a summation over classical orbits of a billiard ball. The summation of orbits is handled by means of the statistical properties of chaotic billiards revealed in Chapter 3, which leads to the universal behavior of conductance in open chaotic quantum dots. A prototype framework of the semiclassical theory is presented here and is used to analyze the quantum phenomena by means of knowledge of classical dynamics. Many themes in the forthcoming chapters spring from the description in this chapter. We shall see how the quantum world will come to be interpreted within the language of the classical world.

4.1 Quantum Billiards with Lead Wires

In the previous chapter, we investigated billiard-ball dynamics. A pair of nearby orbits is found to separate rapidly and to show completely different motions in chaotic billiards. Particles there obeyed classical mechanics. In the microscopic world on the submicron scale, an electron stands for a particle and quantum dots for billiard tables. Here one has to resort to quantum mechanics and the electron obeys the Schrödinger equation. Let us consider the time-independent stationary states. The wavefunction ψ for the electron satisfies time-independent Schrödinger equation

$$\nabla^2 \psi(\boldsymbol{r}) + \frac{2m_e E}{\hbar^2} \psi(\boldsymbol{r}) = 0, \tag{4.1}$$

where $\hbar$ is Planck's constant, and $\boldsymbol{r}$ and m_e are the position and effective mass of the electron, respectively. Dirichlet boundary conditions for ψ are assumed to be satisfied at the boundary walls of the billiard table and lead wires. An example of quantum dots with attached lead wires is given in Fig. 4.1. In this example the direct electronic transmission from lead 1 to lead 2 is prohibited by a central obstacle.

Let the x and y axes be parallel and perpendicular to the leads which have width W. For long-enough leads, the wavefunction that satisfies (4.1) at the leads is given as a product of the parallel component of plane-wave type

$$\frac{1}{\sqrt{2\pi\hbar}} \exp(\pm i k_\parallel x) \tag{4.2}$$

and the perpendicular one that vanishes at the walls of the leads,

41

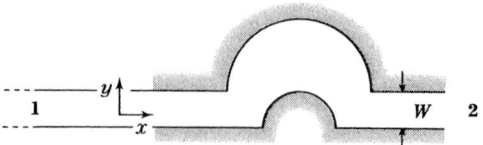

Fig. 4.1   Quantum dot with a pair of lead wires.

$$\phi_m(y) \equiv \sqrt{\frac{2}{W}} \sin\left(\frac{m\pi y}{W}\right). \tag{4.3}$$

$k_\parallel$ is the parallel component of the wave-number vector, and m is a natural number called the **mode number**. m is nothing but the number of half-wavelengths for the standing wave (4.3). The corresponding energy is

$$E = \frac{\hbar^2}{2m_e}\left(k_\parallel^2 + \frac{m^2\pi^2}{W^2}\right), \tag{4.4}$$

from which we find

$$k_\parallel = \pm\sqrt{\frac{2m_e E}{\hbar^2} - \frac{m^2\pi^2}{W^2}} \equiv k_m. \tag{4.5}$$

(4.5) indicates that, for an incident Fermi energy $E = E_{\mathrm{F}}$, the component $k_\parallel$ is also quantized. Thus at the leads the basis wavefunction ψ takes the form

$$\begin{aligned}\psi(x,y) &= \frac{1}{\sqrt{2\pi\hbar}}\exp(\pm ik_\parallel x)\phi_m(y) \\ &= \frac{1}{\sqrt{\pi\hbar W}}\exp\left(\pm ik_m x\right)\sin\left(\frac{m\pi y}{W}\right).\end{aligned} \tag{4.6}$$

The electron represented by the wavefunction (4.6) has momentum vector with parallel component $\pm\hbar k_m$ and perpendicular component $\dfrac{m\pi\hbar}{W}$. Therefore, electronic motion within the leads is oblique against (not parallel to) them: the electron showing a zigzag motion at the leads enters or leaves the billiard table which acts as a cavity. The degree of obliqueness is determined by the mode number m. The schematic zigzag motion at the leads is depicted in Fig. 3.13.

For an open quantum dot as illustrated in Fig. 4.1, the electron with mode m is assumed to come through the left lead 1 into the interior of the billiard table (cavity) and to leave it towards the right lead 2. At the incident lead, the wavefunction that satisfies (4.1) is a sum of the incident and reflecting waves:

$$\psi_1(x,y) = \frac{1}{\sqrt{2\pi\hbar}}e^{ik_m x}\phi_m(y) + \sum_{n=1}^{N_M} S_{nm}^{(1)}\frac{1}{\sqrt{2\pi\hbar}}e^{-ik_n x}\phi_n(y), \tag{4.7}$$

where N_M – the largest integer that does not exceed $\dfrac{\sqrt{2m_e E}\,W}{\pi\hbar}$ – is the total mode number. At the outgoing lead, on the contrary, ψ consists only of the transmission wave:

$$\psi_2(x,y) = \sum_{n=1}^{N_M} S_{nm}^{(2)}\frac{1}{\sqrt{2\pi\hbar}}e^{ik_n x}\phi_n(y). \tag{4.8}$$

ψ_1 and ψ_2 should be smoothly connected via the wavefunction at the billiard table (cavity). This is a basic exercise of quantum mechanics.

$S_{nm}^{(1)}$ in (4.7) and $S_{nm}^{(2)}$ in (4.8) are reflection and transmission components of the S matrix, respectively. They are very important in quantum transport with their values being determined by the boundary condition and dependent on the shapes of the billiard table. According to the Landauer formula, the conductance for a quantum dot is expressed in terms of the transmission components $t_{nm} \equiv \sqrt{\dfrac{k_n}{k_m}}S_{nm}^{(2)}$ as

$$g = \frac{2e^2}{h}T \equiv \frac{2e^2}{h}\sum_{n=1}^{N_M}\sum_{m=1}^{N_M}|t_{nm}|^2, \tag{4.9}$$

where T is the transmission coefficient. An analytic evaluation of t_{nm} is not possible, since the wavefunction at the billiard table is not separable. For an electron with high-enough energy, however, the semiclassical theory can express t_{nm} and T by using the knowledge of classical dynamics in Chapter 3, and eventually lets us understand how the dynamical and statistical features of chaos show up through the conductance.

4.2 Semiclassical Green Function

As shown above, the transmission coefficient T is an important concept in quantum transport. For an electron with high-enough energy so that the corresponding action is much larger than the Planck constant $\hbar$, T can be calculated approximately, and this is the subject of this section.

We start with the semiclassical analysis of the **Green function**. A partner of the Schrödinger equation (4.1) is the equation for the Green function $G(r', r; E)$, i.e.,

$$\nabla_r^2 G(r', r; E) + \frac{2m_e E}{\hbar^2}G(r', r; E) = \frac{2m_e}{\hbar^2}\delta(r - r') \tag{4.10}$$

with Dirichlet boundary condition

$$G(\boldsymbol{r}', \boldsymbol{r}; E) = 0 \tag{4.11}$$

at $\boldsymbol{r}$ and $\boldsymbol{r}'$ on the walls of the billiard table and lead wires. The Green function $G(\boldsymbol{r}', \boldsymbol{r}; E)$ describes the propagation from one state at $\boldsymbol{r}$ to another at $\boldsymbol{r}'$. In terms of the Green function, t_{nm} is expressed by the Fisher–Lee formula[44],

$$t_{nm} = -i\hbar(v_m v_n)^{1/2} \int dy \int dy' \phi_n^*(y') \phi_m(y) G(x', y'; x, y; E).$$

$$\tag{4.12}$$

Here x and x' denote the positions at the points of contact between the cavity and leads, and v_m and v_n the velocity components parallel to the leads. (4.12) tells us that the Green function $G(x', y'; x, y; E)$ determines the transmission coefficient and thereby the conductance in (4.9).

From now on, we shall concentrate on the derivation of the Green function. After defining a **free Green function**, we shall proceed to express the general Green function $G(\boldsymbol{r}', \boldsymbol{r}; E)$ under the Dirichlet boundary condition. The free Green function $G_0(\boldsymbol{r}', \boldsymbol{r}; E)$ – a propagator from $\boldsymbol{r}$ to $\boldsymbol{r}'$ on an infinite plane with neither leads nor billiard table – satisfies (4.10), but is free from the boundary condition (4.11). G_0 is assumed to be propagating as an outward spherical wave, and is expressed in two dimensions in terms of the zeroth order Hankel function of the first kind, as

$$G_0(\boldsymbol{r}', \boldsymbol{r}; E) = -\frac{m_e}{2\hbar^2} i H_0^{(1)} \left(\frac{\sqrt{2m_e E}}{\hbar} |\boldsymbol{r}' - \boldsymbol{r}| \right). \tag{4.13}$$

Indeed $G_0(\boldsymbol{r}', \boldsymbol{r}; E)$ describes direct propagation from $\boldsymbol{r}$ to $\boldsymbol{r}'$ without any bouncing at billiard table walls.

The Green function $G(\boldsymbol{r}', \boldsymbol{r}; E)$ in the presence of a billiard table can be constructed from G_0 by using the **multiple scattering expansion**[14,53]. Suppressing the detailed technique (Harayama *et al.*[53]), the result is as follows:

$$
\begin{aligned}
G\,(\boldsymbol{r}', \boldsymbol{r}; E) \\
= G_0(\boldsymbol{r}', \boldsymbol{r}) - 2 \int ds_1 \left(\frac{\hbar^2}{2m_e} \right) G_0(\boldsymbol{r}', s_1) \frac{\partial G_0(s_1, \boldsymbol{r})}{\partial n_1} \\
+ (-2)^2 \iint ds_1 ds_2 \left(\frac{\hbar^2}{2m_e} \right)^2 G_0(\boldsymbol{r}', s_2) \frac{\partial G_0(s_2, s_1)}{\partial n_2} \frac{\partial G_0(s_1, \boldsymbol{r})}{\partial n_1} \\
+ (-2)^3 \iiint ds_1 ds_2 ds_3 \left(\frac{\hbar^2}{2m_e} \right)^3 \\
\times G_0(\boldsymbol{r}', s_3) \frac{\partial G_0(s_3, s_2)}{\partial n_3} \frac{\partial G_0(s_2, s_1)}{\partial n_2} \frac{\partial G_0(s_1, \boldsymbol{r})}{\partial n_1} \\
+ \cdots + (-2)^N \int \cdots \int ds_1 ds_2 \cdots ds_N \left(\frac{\hbar^2}{2m_e} \right)^N \\
\times G_0(\boldsymbol{r}', s_N) \frac{\partial G_0(s_N, s_{N-1})}{\partial n_N} \cdots \frac{\partial G_0(s_2, s_1)}{\partial n_2} \frac{\partial G_0(s_1, \boldsymbol{r})}{\partial n_1} + \cdots.
\end{aligned}
\tag{4.14}
$$

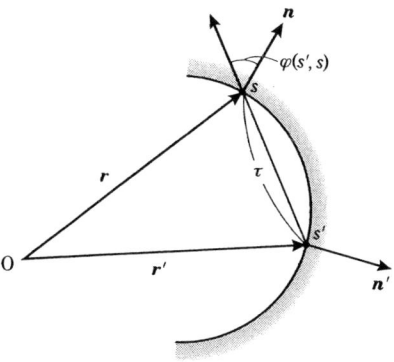

Fig. 4.2

In (4.14) s_i and ds_i are a point on the wall and its small increment along the wall, respectively; $\dfrac{\partial}{\partial n_{s_i}}$ denotes the inward[3] normal derivative at s_i (see Fig. 4.2). The multiple scattering expansion (4.14) looks like the path-integral representation of the Green function to be introduced in the next chapter.

Hereafter, let us investigate (4.14) in the limit of high-enough energy E, i.e., in the semiclassical regime. The inward normal derivative of G_0 in (4.14) is rewritten as

$$\frac{\partial G_0(\boldsymbol{r}',\boldsymbol{r};E)}{\partial n} = \boldsymbol{n} \cdot \nabla_{\boldsymbol{r}} G_0(\boldsymbol{r}',\boldsymbol{r};E)$$

$$= \cos\varphi(s',s)\frac{\partial G_0(\tau)}{\partial \tau}, \tag{4.15}$$

where $\tau\,(\equiv |\boldsymbol{r} - \boldsymbol{r}'|)$ denotes the distance between $\boldsymbol{r}$ and $\boldsymbol{r}'$. As shown in Fig. 4.2, $\varphi(s',s)$ is the angle of the relative position vector $\boldsymbol{r} - \boldsymbol{r}'$ against the inward-normal vector $\boldsymbol{n}$ of the wall with $\varphi(s',s)$ positive for a counter-clockwise rotation of the former. The derivative $\dfrac{\partial G_0}{\partial \tau}$ in (4.15) can be expressed in terms of the first order Hankel function of the first kind, $H_1^{(1)}$, as

$$\frac{\partial G_0}{\partial \tau} = \frac{m_e}{2\hbar^2}\frac{\sqrt{2m_e E}}{\hbar}iH_1^{(1)}\left(\frac{\sqrt{2m_e E}}{\hbar}|\boldsymbol{r} - \boldsymbol{r}'|\right). \tag{4.16}$$

The asymptotic behavior of both G_0 and $\dfrac{\partial G_0}{\partial \tau}$ is straightforward from that of the ν-th order ($\nu = 0, 1$) Hankel function of the first kind[52] at $|z| \gg 1$,

---

[3]This direction implies the one towards the interior of a billiard table wall.

$$H_\nu^{(1)}(z) \approx \sqrt{\frac{2}{\pi z}} e^{i\left(z - (2\nu+1)\frac{\pi}{4}\right)}. \tag{4.17}$$

Substituting (4.13), (4.15) and (4.16) together with the approximation (4.17) into (4.14), we find

$$G(r', r; E)$$

$$\approx G_0(r', r; E) + \sum_{N=1}^{\infty} (-2)^N \left(-\frac{m_e}{2\hbar^2}\right) ie^{-\frac{1}{4}(N+1)\pi i}$$

$$\times \left(\frac{1}{4}\frac{\sqrt{2m_e E}}{\hbar}\right)^N \int \cdots \int ds_1 \cdots ds_N \left(\prod_{j=1}^{N} \cos\varphi(s_{j-1}, s_j)\right)$$

$$\times \left(\prod_{j=1}^{N+1} \sqrt{\frac{2\hbar}{\pi\sqrt{2m_e E}\tau(s_{j-1}, s_j)}}\right) \exp\left(i\frac{\sqrt{2m_e E}}{\hbar}l(s_0, s_1, \cdots, s_{N+1})\right). \tag{4.18}$$

Here s_0 and s_{N+1} stand for the starting r and terminating r' points, respectively, and

$$l(s_0, s_1, \cdots, s_N, s_{N+1}) = \sum_{j=1}^{N+1} \tau(s_{j-1}, s_j)$$

is a whole length of a sequence of linear chords each with length $\tau(s_{j-1}, s_j)$.

A sequence of values, $s_0, s_1, s_2, \cdots, s_{N+1}$, except for both edges, is arbitrary in quantum mechanics, so long as they keep this order along the billiard table wall. Each sequence represents a quantum-mechanical path. (4.18) implies a summation over all possible paths. For an electron with sufficiently high energy, however, the exponential function in the integrand in (4.18) that includes the length l of each path as a phase shows a violent oscillation against a small change of path and is self-destructive. An exception occurs when the phase is stationary and the integrals can be determined by a contribution from the path that gives the stationary phase. This is the spirit of the stationary-phase approximation.

Let us expand the phase – the length of the path – in the exponential function in (4.18) around some path $s_1, \cdots, s_N$ up to term quadratic in the variation δs:

$$l(s_0, s_1 + \delta s_1, s_2 + \delta s_2, \cdots, s_N + \delta s_N, s_{N+1})$$

$$\approx l(s_0, s_1, \cdots, s_N, s_{N+1}) + \sum_{i=1}^{N} \frac{\partial l(s_0, s_1, \cdots, s_N, s_{N+1})}{\partial s_j} \delta s_j$$

$$+ \frac{1}{2} \sum_{i,j=1}^{N} \frac{\partial^2 l(s_0, s_1, \cdots, s_N, s_{N+1})}{\partial s_i \partial s_j} \delta s_i \delta s_j. \tag{4.19}$$

The condition of a stationary phase is given by

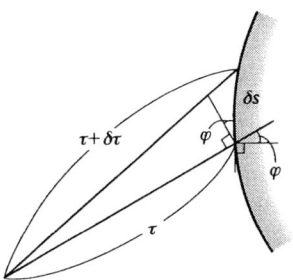

Fig. 4.3

$$\frac{\partial l(s_0, s_1^*, \cdots, s_N^*, s_{N+1})}{\partial s_j} = 0 \qquad (j = 1, 2, \cdots, N), \qquad (4.20)$$

by which the term linear in δs vanishes in (4.19). What path does the sequence $s_0, s_1^*, \cdots, s_N^*, s_{N+1}$, determined by (4.20), represent?

(4.20) is equivalent to the condition for a pair of local chord lengths,

$$\frac{\partial \tau(s_{j-1}^*, s_j^*)}{\partial s_j} + \frac{\partial \tau(s_j^*, s_{j+1}^*)}{\partial s_j} = 0 \qquad (j = 1, 2, \cdots, N). \qquad (4.21)$$

In addition, from Fig. 4.3 we find the relations

$$\frac{\partial \tau(s_{j-1}^*, s_j^*)}{\partial s_j} = \sin \varphi(s_{j-1}^*, s_j^*) \qquad (4.22a)$$

$$\frac{\partial \tau(s_j^*, s_{j+1}^*)}{\partial s_j} = \sin \varphi(s_{j+1}^*, s_j^*). \qquad (4.22b)$$

(4.21) then proves to be nothing but the condition

$$\sin \varphi(s_{j-1}^*, s_j^*) = - \sin \varphi(s_{j+1}^*, s_j^*), \qquad (4.23)$$

which in turn reduces to

$$\varphi(s_{j-1}^*, s_j^*) = -\varphi(s_{j+1}^*, s_j^*) \left(= \varphi(s_j^*, s_{j+1}^*)\right) \qquad (4.24)$$

noting the admissible range of angles $\varphi\left(-\frac{\pi}{2} \leq \varphi \leq \frac{\pi}{2}\right)$.

(4.23) and (4.24) are just the **law of reflection** that the angle of incidence agrees with that of reflection. Therefore the path characterized by the sequence $s_0, s_1^*, \cdots, s_N^*, s_{N+1}$, is the one that starts at $\boldsymbol{r}\ (= s_0)$, satisfies the law of reflection at $s_1^*, \cdots, s_N^*$ on the wall, and reaches $\boldsymbol{r}'(= s_{N+1})$. Such a path is naturally the orbit of a billiard ball mentioned in chapter 3. In this way, for an electron with high-enough energy so that its corresponding action $\gg \hbar$, the path satisfying the law of reflection has been selected among the infinitely large number of paths admissible in quantum mechanics, namely, quantum mechanics has come to be described approximately by using the information in classical mechanics.

We now proceed to investigate the quantum effect. In (4.19), the condition of the stationary phase sweeps away the terms linear in $\delta s_1, \cdots, \delta s_N$ but left those quadratic in $\delta s_1, \cdots, \delta s_N$. The latter are called quantum fluctuations, which will be incorporated into (4.18) in the semiclassical approximation (or the stationary-phase approximation) in the following way.

The transformation of the principal axis reduces the term quadratic in $\delta s_1, \cdots, \delta s_N$ to the diagonal form like $\sum\limits_{i=1}^{N} A_i \delta s_i^2$. Then N independent Gaussian integrations lead to the factor $\prod\limits_{i=1}^{N} \sqrt{\dfrac{\pi}{|A_i|}}$. Further, the product $\prod\limits_{i=1}^{N} A_i$ can be identified with the determinant of the matrix responsible for the quadratic form before the transformation of principal axis. Applying these arguments, (4.18) becomes:

$$
\begin{aligned}
G(r', r; E) \\
\approx G_0(r, r'; E) + \sum_{N=1}^{\infty} \sum_{\text{Orbits}} (-1)^N \left(-\frac{1}{4}i\right) \left(\frac{2m_e}{\hbar^2}\right) \sqrt{\frac{2\hbar}{\pi \sqrt{2m_e E}}} \\
\times \left[\frac{\prod\limits_{j=1}^{N} \cos \varphi(s_{j-1}^*, s_j^*)}{\prod\limits_{i=1}^{N+1} \sqrt{\tau(s_{j-1}^*, s_j^*)}} \right] \frac{1}{\sqrt{|\det W|}} \\
\times \exp\left(i \frac{\sqrt{2m_e E}}{\hbar} l(s_0, s_1^*, \cdots, s_N^*, s_{N+1}) - \frac{\sigma}{2}\pi i - \frac{1}{4}\pi i \right),
\end{aligned} \tag{4.25}
$$

where $\sum\limits_{\text{Orbits}}$ means the summation over all orbits that collide with the billiard table wall N times and satisfy the law of reflection at the collision points $s_1^*, \cdots, s_N^*$; W is the $N \times N$ matrix with elements given by

$$
W_{ij} = \frac{\partial^2 l}{\partial s_i \partial s_j}(s_0, s_1^*, \cdots, s_N^*, s_{N+1}) \qquad (i, j = 1, 2, \cdots, N), \tag{4.26}
$$

and

$$
\sigma = \text{number of negative eigenvalues}^4 \text{ for } W \tag{4.27}
$$

is the compensation for taking the absolute value of $\det W$ in (4.25). Note: throughout the semiclassical approximation leading to (4.25) we assumed each path $s_1^*, \cdots, s_N^*$ is isolated (not degenerate). This assumption and therefore the semiclassical approximation are guaranteed in chaotic billiards.

---

[4] In general σ is the number of self-conjugate points of each orbit.

We shall now rewrite the pre-exponential factor in each term in (4.25) to make its physical meaning crystal-clear. For this purpose, let us prove the following relation,

$$\frac{\displaystyle\prod_{i=1}^{N+1}\left(-\frac{\partial^2 l}{\partial s_{j-1}\partial s_j}(s_0, s_1^*, \cdots, s_N^*, s_{N+1})\right)}{\det \boldsymbol{W}}$$
$$= \frac{1}{(M_{N+1\leftarrow N}M_{N\leftarrow N-1}\cdots M_{2\leftarrow 1}M_{1\leftarrow 0})_{12}}, \qquad (4.28)$$

where $M_{i+1\leftarrow i}$ is the 2×2 transformation matrix for increments of the Birkhoff coordinates in Section 3.2, and $(M_{N+1\leftarrow N}M_{N\leftarrow N-1}\cdots M_{2\leftarrow 1}M_{1\leftarrow 0})$ tells us how the initial increments in position and direction of emission propagate due to successive collisions of a billiard ball with the wall.

First of all, under the fixed value for both ends (s_0 and s_{N+1}) of the path, let us take the variation of the stationary-phase condition (4.21) as

$$\frac{\partial^2 \tau(s_{i-1}, s_i)}{\partial s_{i-1}\partial s_i}\delta s_{i-1} + \left(\frac{\partial^2 \tau(s_{i-1}, s_i)}{\partial s_i^2} + \frac{\partial^2 \tau(s_i, s_{i+1})}{\partial s_i^2}\right)\delta s_i$$
$$+ \frac{\partial^2 \tau(s_i, s_{i+1})}{\partial s_{i+1}\partial s_i}\delta s_{i+1} = 0$$
$$(i = 1, 2, \cdots, N) \qquad (4.29)$$

with $\delta s_0 = \delta s_{N+1} = 0$, which provides a law obeyed by deviations from the classical path ($\delta s_1, \cdots, \delta s_N$). Equation (4.29) can be rewritten in terms of the transformation matrices for the linearized Birkhoff map in (3.31). Following (3.31), the increments δs_i in position and $\delta \varphi_i$ in angle of emission just after the i-th collision will be transformed via the matrix $M_{i+1\leftarrow i}$ as

$$\begin{pmatrix} \delta s_{i+1} \\ \delta(\sin \varphi_{i+1}) \end{pmatrix} = M_{i+1\leftarrow i}\begin{pmatrix} \delta s_i \\ \delta(\sin \varphi_i) \end{pmatrix}. \qquad (4.30)$$

Noting the initial condition $\delta s_0 = 0$, (4.30) yields

$$\delta s_1 = (M_{1\leftarrow 0})_{12}\,\delta(\sin \varphi_0), \qquad (4.31a)$$
$$\delta s_{N+1} = (M_{N+1\leftarrow N}M_{N\leftarrow N-1}\cdots M_{1\leftarrow 0})_{12}\,\delta(\sin \varphi_0), \qquad (4.31b)$$

and their combination

$$\delta s_{N+1} = \frac{(M_{N+1\leftarrow N}\cdots M_{1\leftarrow 0})_{12}}{(M_{1\leftarrow 0})_{12}}\delta s_1. \qquad (4.32)$$

The equivalence between (4.20) and (4.21) together with (4.32) reduces (4.29) to the matrix form,

$$L\delta \boldsymbol{s} = 0, \qquad (4.33)$$

where L is a $N \times N$ matrix given by

$$L \equiv \begin{pmatrix} \dfrac{\partial^2 l}{\partial s_1^2} & \dfrac{\partial^2 l}{\partial s_1 \partial s_2} & 0 & & & \\[2mm] \dfrac{\partial^2 l}{\partial s_2 \partial s_1} & \dfrac{\partial^2 l}{\partial s_2^2} & \dfrac{\partial^2 l}{\partial s_2 \partial s_3} & & & \\[2mm] 0 & \dfrac{\partial^2 l}{\partial s_3 \partial s_2} & \dfrac{\partial^2 l}{\partial s_3^2} & & & \\[2mm] \vdots & & \cdots & & & \\[2mm] \vdots & & & & & \\[2mm] 0 & & \cdots & \cdots & & \\[2mm] \dfrac{(M_{N+1 \leftarrow N} \cdots M_{1 \leftarrow 0})_{12}}{(M_{1 \leftarrow 0})_{12}} \dfrac{\partial^2 l}{\partial s_N \partial s_{N+1}} & 0 & \cdots & & & \end{pmatrix}$$

$$\begin{pmatrix} \cdots & \cdots & 0 & 0 \\[2mm] 0 & \cdots & \cdots & 0 \\[2mm] \dfrac{\partial^2 l}{\partial s_3 \partial s_4} & 0 & \cdots & 0 \\[2mm] \cdots & \cdots & & \vdots \\[2mm] \cdots & \cdots & \cdots & 0 \\[2mm] 0 & \dfrac{\partial^2 l}{\partial s_{N-1} \partial s_{N-2}} & \dfrac{\partial^2 l}{\partial s_{N-1}^2} & \dfrac{\partial^2 l}{\partial s_{N-1} \partial s_N} \\[2mm] \cdots & 0 & \dfrac{\partial^2 l}{\partial s_N \partial s_{N-1}} & \dfrac{\partial^2 l}{\partial s_N^2} \end{pmatrix} \qquad (4.34)$$

and δs is an N-component vector, i.e.,

$$\delta s \equiv \begin{pmatrix} \delta s_1 \\ \delta s_2 \\ \vdots \\ \delta s_{N-1} \\ \delta s_N \end{pmatrix}. \qquad (4.35)$$

The condition for (4.33) to have a nontrivial solution is

$$\det L = 0. \qquad (4.36)$$

Using the definition (4.26), the Laplace expansion of (4.36) with respect to the first column in (4.34) leads to

$$\det \boldsymbol{W} - \frac{(M_{N+1 \leftarrow N} \cdots M_{1 \leftarrow 0})_{12}}{(M_{1 \leftarrow 0})_{12}} \prod_{i=1}^{N} \left(-\frac{\partial^2 l}{\partial s_i \partial s_{i+1}} \right) = 0. \qquad (4.37)$$

There is a supplemental relation obtained from (4.31a),

$$\frac{1}{(M_{1\leftarrow 0})_{12}} = \left(\frac{\partial \sin \varphi_0(s_0, s_1)}{\partial s_1}\right)_{s_0} = -\frac{\partial^2 l}{\partial s_0 \partial s_1}. \tag{4.38}$$

A combination of (4.37) and (4.38) yields (4.28), thereby completing the proof.

In the interesting equality (4.28), factors other than $\det \boldsymbol{W}$ can further be rewritten. First, the derivative of (4.22a) and (4.22b) gives the transformation

$$\frac{\partial^2 l}{\partial s_{j-1} \partial s_j}(s_0, s_1^*, \cdots, s_N^*, s_{N+1}) = \frac{\cos \varphi(s_{j-1}, s_j) \cos \varphi(s_j, s_{j+1})}{\tau(s_{j-1}, s_j)}, \tag{4.39}$$

and second, (4.31b) gives the reduction

$$\frac{1}{(M_{N+1\leftarrow N} M_{N\leftarrow N-1} \cdots M_{2\leftarrow 1} M_{1\leftarrow 0})_{12}} = \cos \varphi_0 \frac{\partial \varphi_0}{\partial s_{N+1}}. \tag{4.40}$$

Consequently, (4.28) combined with (4.39) and (4.40) leads to the very helpful relation,

$$\left[\frac{\prod_{j=1}^{N} \cos \varphi(s_{j-1}^*, s_j^*)}{\prod_{i=1}^{N+1} \sqrt{\tau(s_{j-1}^*, s_j^*)}}\right] \frac{1}{\sqrt{|\det \boldsymbol{W}|}} = \left[\frac{1}{\cos \varphi_{N+1}} \left|\frac{\partial \varphi_0}{\partial s_{N+1}}\right|\right]^{1/2}. \tag{4.41}$$

The right-hand side of the above equality is expressed in terms of the ratio between the initial and final increments in the classical path of a billiard ball.

The left-hand side of (4.41) constitutes the main part of the pre-exponential factor in the expansion (4.25): this expansion with the use of (4.41) now reads

$$G(\boldsymbol{r}', \boldsymbol{r}; E)$$

$$\approx G_0(\boldsymbol{r}', \boldsymbol{r}; E) + \sum_{\text{Orbit:}\gamma} \left(-\frac{1}{4}i\right) \left(\frac{2m_e}{\hbar^2}\right) \sqrt{\frac{2\hbar}{\pi m_e v}} \left[\frac{1}{\cos \varphi_{N+1}} \left|\frac{\partial \varphi_0}{\partial s_{N+1}}\right|\right]^{1/2}$$

$$\times \exp \left(i \frac{\sqrt{2m_e E}}{\hbar} l_\gamma - \frac{\sigma_\gamma}{2}\pi i - \frac{1}{4}\pi i + N_\gamma \pi i\right)$$

$$= G_0(\boldsymbol{r}', \boldsymbol{r}; E) - i e^{-\frac{1}{4}\pi i} \sum_{\text{Orbit:}\gamma} \frac{2\pi}{(2\pi\hbar)^{3/2}} \left[\frac{m_e}{v \cos \varphi_{N+1}} \left|\frac{\partial \varphi_0}{\partial s_{N+1}}\right|\right]^{1/2}$$

$$\times \exp \left(i \frac{\sqrt{2m_e E}}{\hbar} l_\gamma - \frac{\sigma_\gamma}{2}\pi i + N_\gamma \pi i\right). \tag{4.42}$$

The Green function is thus expressed as a summation over classical paths: $\sum_{\text{Orbit:}\gamma}$ denotes a sum over all orbits (now free from the number of bouncing with wall boundaries) coming into the billiard system at $\boldsymbol{r}$ and leaving it at $\boldsymbol{r}'$; N_γ and σ_γ are the number of collisions with walls and that of caustics for each orbit γ, respectively; φ_0

and φ_{N+1} measured at r and r', respectively, represent the angles of the direction of emission against the inwardly normal direction of the wall with positive angles for clockwise rotation (see Fig. 4.3); s_{N+1} is the length between a terminal r' and a suitable origin on the wall measured along the wall boundary.

The free Green function $G_0(r', r; E)$ on the first term in the above expansion can also be expressed on a similar footing: from (4.13) and (4.17), its semiclassical limit is given by

$$G_0(r', r; E) \approx -ie^{-\frac{1}{4}\pi i} \frac{2\pi}{(2\pi\hbar)^{3/2}} \left[\frac{m_e}{v} \frac{1}{|r - r'|} \right]^{1/2}$$
$$\times \exp\left(i \frac{\sqrt{2m_e E}}{\hbar} |r - r'| \right), \qquad (4.43)$$

which together with

$$\frac{\partial \varphi_0}{\partial s_{N+1}} = \frac{1}{|r - r'|}$$

gives the same form as those of the remaining terms in (4.42), and is taken as an extra contribution due to the direct path connecting r and r'. Incorporating (4.43) into the sum over orbits, we arrive at

$$G(r', r; E) \approx G_{\text{scl}}(r', r; E)$$
$$\equiv -ie^{-\frac{1}{4}\pi i} \sum_{\text{Orbit}:\gamma} \frac{2\pi}{(2\pi\hbar)^{3/2}} \left[\frac{m_e}{v \cos \varphi_{N+1}} \left| \frac{\partial \varphi_0}{\partial s_{N+1}} \right| \right]^{1/2}$$
$$\times \exp\left(i \frac{\sqrt{2m_e E}}{\hbar} l_\gamma - \frac{\sigma_\gamma}{2}\pi i + N_\gamma \pi i \right). \qquad (4.44)$$

This form shows that the semiclassical approximation makes it possible to describe the Green function for billiard systems as a sum of contributions over classical paths of a billiard ball. Recalling the fundamental role of the Green function in quantum physics, various quantum phenomena have come to be interpreted on the basis of the underlying classical paths.

In strongly chaotic systems where each orbit shows an extreme sensitivity to initial conditions, individual orbits connecting r and r' and satisfying the law of reflection between them are isolated from each other. Equation (4.44) is the semiclassical result for such a strongly chaotic case. In systems with continuous symmetry such as circle billiards, orbits that satisfy the law of reflection are not isolated and there exist other orbits infinitesimally close to a given orbit. The semiclassical expression for such infinitely degenerate cases will be mentioned briefly in the forthcoming chapters.

4.3   Transmission Coefficients

As explained in Section 4.1, electrical conductance is obtained from transmission coefficients, which in turn is derived from the Green function. In Section 4.2, we learnt that in the semiclassical limit, the Green function is calculable as a sum over classical orbits of a billiard ball. In this section, this **semiclassical Green function** will be applied to derive the transmission coefficients T. The description below follows the scheme of Jalabert *et al.*[45-47].

Let us calculate the transmission components of the S matrix, t_{nm}, for a quantum dot with leads as in Fig. 4.1. The point of incidence $r = (x, y)$ and that of ejection $r' = (x', y')$ are defined at the point of contact $(x = 0)$ of the billiard table (cavity) with the left lead and that $(x = L)$ with the right lead, respectively. At $x = 0$, $s_0 = y$ is measured perpendicularly to the lead and φ_0 is the angle between the direction of emission and the x axis. Similarly at $x' = L$, $s_{N+1} = y'$ and $\varphi_{N+1} = \theta'$. To summarize,

$$t_{nm} \approx -i\hbar(v_m v_n)^{1/2} \int dy' \int dy \phi_n^*(y')\phi_m(y)G_{\text{scl}}(L, y'; 0, y; E)$$

(4.45)

with

$$G_{\text{scl}}(L, y'; 0, y; E) = -ie^{-\frac{1}{4}\pi i} \sum_{\text{Orbit}:\gamma} \frac{2\pi}{(2\pi\hbar)^{3/2}} \left[\frac{m_e}{v\cos\theta'} \left| \frac{\partial\theta}{\partial y'} \right| \right]^{1/2}$$
$$\times \exp\left(i\frac{\sqrt{2m_e E}}{\hbar} l_\gamma - \frac{\sigma_\gamma}{2}\pi i + N_\gamma \pi i \right).$$

(4.46)

v_m and v_n denote parallel velocity components at the incident and outgoing leads, respectively.

For electrons with high-enough energy, the integration over variables y and y' in (4.45) is evaluated as before in the stationary-phase approximation. Using the wavefunctions at the leads,

$$\phi_m(y) \equiv \sqrt{\frac{2}{W}} \sin\left(\frac{m\pi y}{W}\right)$$
$$= \sqrt{\frac{2}{W}} \frac{1}{2i} \left(\exp\left(i\frac{m\pi y}{W}\right) - \exp\left(-i\frac{m\pi y}{W}\right) \right),$$

(4.47)

(4.45) can be shown to possess the phase part in each exponential function:

$$\frac{\sqrt{2m_e E}}{\hbar} l_\gamma \pm m\frac{\pi y}{W} \pm n\frac{\pi y'}{W}.$$

The phase stationarity with respect to y is then satisfied for

$$\frac{\sqrt{2m_e E}}{\hbar} \frac{\partial l_\gamma}{\partial y} + \frac{\overline{m}\pi}{W} = 0$$

(4.48)

with $\overline{m} \equiv \pm m$. Noting the equality

$$\frac{\partial l_\gamma}{\partial y} = -\sin\theta \tag{4.49}$$

(see (4.22b)), the condition (4.48) says that the initial momentum component $p_y = \sqrt{2m_e E}\sin\theta$ should satisfy

$$p_y = \frac{\overline{m}\hbar\pi}{W}. \tag{4.50}$$

Similarly, the phase stationarity with respect to y' asserts that the momentum component at the point of ejection $p'_y = \sqrt{2m_e E}\sin\theta$ satisfies

$$p'_y = \sqrt{2m_e E}\sin\theta' = -\frac{\overline{n}\hbar\pi}{W}. \tag{4.51}$$

As understood from (4.50) and (4.51), for a given perpendicular wavefunction $\phi_m(y)$ $(\phi_n(y'))$ at the entrance (exit) lead, there exist classical orbits of a billiard ball with initial angle of emission and final angle of ejection defined by the mode numbers m and n, respectively. A sum of contributions from those orbits gives the semiclassical value of t_{nm}.

Let y_0 and y'_0 be the location for emission and that of ejection, respectively, which guarantee the phase stationarity, and define the effective action

$$\widetilde{S}(\overline{m},\overline{n};E) \equiv \sqrt{2m_e E}l(y_0,y'_0) + \frac{\hbar\pi\overline{m}}{W}y_0 - \frac{\hbar\pi\overline{n}}{W}y_{0'}. \tag{4.52}$$

In terms of $\widetilde{S}$, t_{nm} reads

$$t_{nm} = -e^{\frac{\pi i}{4}}\frac{\sqrt{2\pi\hbar}}{2W}\sum_{s(\overline{n},\overline{m})}\mathrm{sgn}(\overline{m})\,\mathrm{sgn}(\overline{n})\sqrt{D_s}\exp\left(\frac{i}{\hbar}\widetilde{S}_s(\overline{m},\overline{n};E) - i\frac{\pi}{2}\widetilde{\mu}_s\right), \tag{4.53}$$

where $\sum_{s(\overline{n},\overline{m})}$ means summation over all orbits $s(\overline{n},\overline{m})$ that simultaneously satisfy (4.50) and (4.51); $\mathrm{sgn}(a)$ denotes the sign of a; the index $\widetilde{\mu}$, including the number of caustics and that of collisions, is defined as follows:

$$\widetilde{\mu} = \sigma - 2N + H\left(-\left(\frac{\partial\theta}{\partial y}\right)_{y'}\right) + H\left(-\left(\frac{\partial\theta'}{\partial y'}\right)_y\right), \tag{4.54}$$

where H stands for the Heaviside step function[5] and D_s is

$$D_s \equiv \frac{1}{m_e v\cos\theta'}\left|\left(\frac{\partial y}{\partial\theta'}\right)_\theta\right|.$$

In the argument so far, we have assumed no magnetic field. In the presence of a weak magnetic field, a perturbational treatment is possible: the momentum acquires

[5] $H(t) = \begin{cases} 0 & (t < 0) \\ 1 & (t \geq 0). \end{cases}$ Although $\Theta(t)$ is sometimes used, we use the present symbol so as not to confuse it with angle or area variables.

an extra term $\dfrac{e}{c}\boldsymbol{A}$ due to the vector potential $\boldsymbol{A}$. By simply adding to $\widetilde{S}(\overline{m},\overline{n};E)$ in (4.53) the integration of this term along the classical orbit, one will find an improved t_{nm} in the presence of a weak magnetic field.

The semiclassical transmission coefficient T is a sum of $|t_{nm}|^2$ over all possible mode numbers ($m = 1, \cdots, N_M$ and $n = 1, \cdots, N_M$) and is summarized as

$$T(k) = \sum_{n=1}^{N_M}\sum_{m=1}^{N_M} |t_{nm}|^2 = \frac{1}{2}\frac{\pi}{kW}\sum_{n=1}^{N_M}\sum_{m=1}^{N_M}\sum_{s}\sum_{u} F_{n,m}^{s,u}(k), \qquad (4.55)$$

where

$$F_{n,m}^{s,u}(k) = \sqrt{\widetilde{A}_s \widetilde{A}_u}\exp\left[ik(\widetilde{L}_s - \widetilde{L}_u) + i\pi\phi_{s,u}\right]. \qquad (4.56)$$

Here both s and u are the orbits satisfying (4.50) and (4.51) at the same time; $\widetilde{L}_s = \dfrac{\widetilde{S}_s(\overline{m},\overline{n})}{k\hbar}$ is the effective orbital length and $\phi_{s,u} = \dfrac{1}{2}(\tilde{\mu}_u - \tilde{\mu}_s)$. The stability factor is now prescribed as $\widetilde{A}_s = \dfrac{\hbar k}{W}\widetilde{D}_s$.

We are now ready to proceed to the conductance given by the Landauer formula. In the final section, we shall show how the statistical features of billiard ball dynamics elucidated in Chapter 3 will affect the conductance fluctuations.

4.4 Conductance Fluctuations

Let us embark upon the ultimate stage to resolve the problem raised in Chapter 2, namely, that of evaluating semiclassically the autocorrelation function for the **conductance fluctuations** as a function of the magnetic field increment:

$$C(\Delta B) \equiv \langle \delta T(k, B + \Delta B)\delta T(k, B)\rangle_k, \qquad (4.57)$$

where $\langle \cdots \rangle_k$ means the average over wave number (or energy),

$$\langle \cdots \rangle_k = \lim_{q\to\infty}\frac{1}{q}\int_{k}^{k+q} dk', \qquad (4.58)$$

and $\delta T(k, B)$ is the fluctuation of transmission coefficient $T(k, B)$ around its mean,

$$\delta T(k, B) = T(k, B) - \langle T(k, B)\rangle_k. \qquad (4.59)$$

Firstly, we shall investigate the mean transmission coefficient in the large wave-number limit (the semiclassical limit). $\langle T(k, B)\rangle_k$ is expected to be proportional to the total mode number $N_M = \left[\dfrac{kW}{\pi}\right] \approx \dfrac{kW}{\pi}$. Therefore, bringing in the prescription

$$\langle T(k, B)\rangle_k = \frac{kW}{\pi} t_{\mathrm{cl}}, \tag{4.60}$$

let us define

$$t_{\mathrm{cl}} \equiv \left\langle \frac{\pi T(k, B)}{kW} \right\rangle_k, \tag{4.61}$$

which, combined with (4.55), gives

$$t_{\mathrm{cl}} = \frac{1}{2}\left(\frac{\pi}{kW}\right)^2 \sum_{n=1}^{N_M} \sum_{m=1}^{N_M} \sum_s \sum_u \lim_{q\to\infty} \int_k^{k+q} \frac{1}{q}\sqrt{\tilde{A}_s \tilde{A}_u}$$
$$\times \exp\left[ik'(\tilde{L}_s - \tilde{L}_u) + i\pi\phi_{s,u}\right] dk'. \tag{4.62}$$

In the above double summation over orbits s and u, only the diagonal term with $s = u$ survives in the stationary-phase approximation. Hence,

$$t_{\mathrm{cl}} = \frac{1}{2}\left(\frac{\pi}{kW}\right)^2 \sum_{n=-N_M}^{n=N_M} \sum_{m=-N_M}^{n=N_M} \sum_s \tilde{A}_s. \tag{4.63}$$

Since the incidence angle is

$$\sin\theta_m \equiv \frac{m}{\left(\dfrac{kW}{\pi}\right)},$$

we take the continuum limit

$$\frac{\pi}{kW} \sum_{m=-N_M}^{N_M} f(\theta_m) \longrightarrow \int_{-1}^{1} d(\sin\theta) f(\theta).$$

In this limit, (4.63) is reduced to

$$t_{\mathrm{cl}} = \frac{1}{2}\int_{-1}^{1} d(\sin\theta) \int_{-1}^{1} d(\sin\theta') \sum_{s(\theta,\theta')} \tilde{A}_s$$
$$= \frac{1}{2}\int_{-1}^{1} d(\sin\theta) \int_0^W \frac{dy}{W} f(y, \theta), \tag{4.64}$$

where

$$\tilde{A}_s = \frac{\hbar k}{W}\tilde{D}_s = \frac{1}{W|\cos\theta'|}\left|\left(\frac{\partial y}{\partial\theta'}\right)_\theta\right| \tag{4.65}$$

has been substituted. $f(y, \theta)$ in (4.64) is unity, provided a particle with initial position y and incidence angle θ at lead 1 reaches 2, and zero otherwise. As explained in Chapter 3, y and $\sin\theta$ constitute Birkhoff coordinates. The integrals in (4.64) tell us how big a fraction can contribute to the transmission among the initial points uniformly distributed over the $(0, W) \times (-1, 1)$ phase space. Therefore t_{cl} can be interpreted as the **classical transmission coefficient**.

Let us proceed to calculate the autocorrelation function. Contrary to the mean part, fluctuations δT come from contributions due to interference between paths $s \neq u$. From (4.55),

$$
C(\Delta B)
$$
$$
= \frac{1}{4} \left\langle \left(\frac{\pi}{kW} \right)^2 \sum_{n=1}^{N_M} \sum_{m=1}^{N_M} \sum_{n'=1}^{N_M} \sum_{m'=1}^{N_M} \sum_{s,u(u \neq s)} \sum_{t,v(v \neq t)} F_{n,m}^{s,u}(B + \Delta B) F_{n',m'}^{t,v}(B) \right\rangle_k.
$$
(4.66)

We assume that among the summations over modes, terms other than those that satisfy simultaneously $n' = n$ and $m' = m$ vanish in the large wave-number limit. In this diagonal approximation, we have

$$
C(\Delta B) \approx C_D(\Delta B)
$$
$$
= \frac{1}{4} \left\langle \left(\frac{\pi}{kW} \right)^2 \sum_{n=1}^{N_M} \sum_{m=1}^{N_M} \sum_{s,u(u \neq s)} \sum_{t,v(v \neq t)} F_{n,m}^{s,u}(B + \Delta B) F_{n,m}^{t,v}(B) \right\rangle_k.
$$
(4.67)

Replacing the summation over n and m by the integration above (4.64), the integrand in (4.67) proves to have the phase factor,

$$
\sqrt{\tilde{A}_s \tilde{A}_u \tilde{A}_t \tilde{A}_v} \exp\left[ik \left(\tilde{L}_s(B + \Delta B) - \tilde{L}_u(B + \Delta B) + \tilde{L}_t(B) - \tilde{L}_v(B) \right) \right].
$$
(4.68)

Note that so long as the magnetic field B is weak enough, classical orbits are almost straight except for the bouncing points where the law of reflection is satisfied. However, with use of the vector potential $\mathbf{A}$ $(\mathbf{B} = \nabla \times \mathbf{A})$, a new definition of momentum,

$$
\mathbf{p} = m_e \mathbf{v} - \frac{e}{c} \mathbf{A},
$$

should be employed (see the footnote in Section 6.1), which demands a suitable correction in the action.

In averaging over wave numbers (see (4.58)), the phase factors become self-destructive except for the terms that satisfy $v = s$ and $t = u$ at the same time. Other possible interference terms are prohibited because of the existing constraint $s \neq u$ and $t \neq v$ in (4.67). For a pair of paths $v = s$, the effect of the B field is handled perturbatively. Their contribution to the phase in (4.68) is given by

$$
k \left(\tilde{L}_s(B + \Delta B) - \tilde{L}_s(B) \right) = -\frac{e}{\hbar c} \int \Delta \mathbf{A} \cdot d\mathbf{l}
$$

$$= -\frac{e}{\hbar c} \iint (\nabla \times \Delta \boldsymbol{A}) \cdot d\boldsymbol{a}, \tag{4.69}$$

where $\Delta \boldsymbol{A}$ is an increment of $\boldsymbol{A}$. The differential elements $d\boldsymbol{l}$ and $d\boldsymbol{a}$ are defined along the path and for the area enclosed by the path, respectively. Application of Stokes' theorem to the above integral leads to

$$k \left(\widetilde{L}_s(B + \Delta B) - \widetilde{L}_s(B) \right) = \frac{\Delta B}{\Phi_0} \Theta_s \tag{4.70}$$

with $\Delta B = |\nabla \times \Delta \boldsymbol{A}|$ and $\Phi_0 \equiv \dfrac{hc}{e}$; Θ_s is the area ($\times\, 2\pi$) enclosed by orbit s, which is taken positive for clockwise rotation. By taking together another contribution from a pair of paths $t = u$, the autocorrelation function (4.67) is reduced to

$$C_D(\Delta B) = \frac{1}{4} \int_{-1}^{1} d(\sin\theta) \int_{-1}^{1} d(\sin\theta')$$
$$\times \sum_{s(\theta,\theta')} \sum_{u(\theta,\theta')(u \neq s)} \widetilde{A}_s \widetilde{A}_u \exp\left[i\left(\Theta_s - \Theta_u\right) \frac{\Delta B}{\Phi_0} \right]. \tag{4.71}$$

The power spectrum for conductance fluctuations is derived by taking the inverse Fourier transformation of $C_D(\Delta B)$ as

$$\widehat{C}_D(\eta) \equiv \int d(\Delta B) C(\Delta B) \exp\left(i \frac{\eta}{\Phi_0} \Delta B \right)$$
$$= \frac{\Phi_0 \pi}{2} \int_{-1}^{1} d(\sin\theta) \int_{-1}^{1} d(\sin\theta')$$
$$\times \sum_{s(\theta,\theta')} \sum_{u(\theta,\theta')(u \neq s)} \widetilde{A}_s \widetilde{A}_u \delta(\Theta_s + \eta - \Theta_u). \tag{4.72}$$

Substitution of the equality

$$\delta(\Theta_s + \eta - \Theta_u) = \int_{-\infty}^{\infty} d\Theta \delta(\Theta_s + \eta - \Theta)\delta(\Theta - \Theta_u) \tag{4.73}$$

into (4.72) results in

$$\widehat{C}_D(\eta) = \frac{\Phi_0 \pi}{2} \int_{-\infty}^{\infty} d\Theta \int_{-1}^{1} d(\sin\theta_s) \int_{0}^{W} \frac{dy_s}{W} f(y_s, \theta_s)\delta(\Theta_s + \eta - \Theta)$$
$$\times \int_{-1}^{1} d(\sin\theta_u) \int_{0}^{W} \frac{dy_u}{W} f(y_u, \theta_u)\delta(\Theta - \Theta_u). \tag{4.74}$$

The meaning of $f(y,\theta)$ was already given below (4.65). Define the **area distribution**

$$P(\Theta) \equiv \int_{-1}^{1} d(\sin\theta) \int_{0}^{W} \frac{dy}{W} f(y,\theta)\delta(\Theta - \Theta_u), \tag{4.75}$$

which describes how big a fraction of the initial points is uniformly distributed over the $(0, W) \times (-1, 1)$ phase space to depict a given area Θ. Using (4.75) in (4.74) yields the form of convolution

$$\widehat{C}_D(\eta) = \frac{\Phi_0 \pi}{2} \int_{-\infty}^{\infty} d\Theta P(\Theta + \eta)P(\Theta). \tag{4.76}$$

Combining this result with the exponential area distribution for fully chaotic billiards (see the final issue in Chapter 3)

$$P(\Theta) \propto \exp(-\alpha_{\mathrm{cl}}|\Theta|) \tag{4.77}$$

gives rise to a universal power spectrum for conductance fluctuations:

$$\widehat{C}_D(\eta) \propto \int_{-\infty}^{\infty} d\Theta \exp(-\alpha_{\mathrm{cl}}|\Theta + \eta| - \alpha_{\mathrm{cl}}|\Theta|)$$
$$\propto (1 + \alpha_{\mathrm{cl}}|\eta|)\exp(-\alpha_{\mathrm{cl}}|\eta|). \tag{4.78}$$

Now it has become crystal-clear why the power spectrum for conductance fluctuations shows an exponential decay in the case of chaotic billiards.

Noting that the Fourier transformation of $P(\Theta)$ gives $\dfrac{1}{1+\left(\dfrac{\Delta B}{\alpha_{\mathrm{cl}}\Phi_0}\right)^2}$ and applying the theorem of convolution, the Fourier transformation of (4.76) leads to (Jalabert *et al.*[45]),

$$C_D(\Delta B) = \frac{C_D(0)}{\left[1 + \left(\dfrac{\Delta B}{\alpha_{\mathrm{cl}}\Phi_0}\right)^2\right]^2}. \tag{4.79}$$

Thus the autocorelation function for conductance fluctuations is proportional to the square of a Lorentzian function.

The fundamental question raised in Chapter 1 has come to be fully understood: we have answered the question of why quantum dots and classical billiards are classified in a similar way according to their shapes despite the difference of the world in which they exist: microscopic and quantum-mechanical on one side; macroscopic and classical-mechanical on the other. Chaos is a concept of classical dynamics in the macroscopic world, and the area of a billiard ball enclosed until it leaves the chaotic billiard system obeys the exponential law. Electrical conductance for nanoscale quantum dots is concerned with electron dynamics in the microscopic quantum world. However, the semiclassical theory bridges the two different worlds, and has derived an exponential distribution for the power spectrum of conductance fluctuations as a quantum manifestation of classical chaos. Although some of the latest experimental and numerical data indicate the deviation from the exponential law, the framework so far plays a fundamental role in our understanding the quantum transport in the semiclassical regime. In the following chapters, we shall develop the present scheme in several directions and explain many interesting themes on quantum chaos in quantum dots.

5

SEMICLASSICAL QUANTIZATION AND THERMODYNAMICS OF MESOSCOPIC SYSTEMS

In this chapter, to construct thermodynamics of mesoscopic systems, we begin with an investigation of the semiclassical theory for more general Hamiltonian systems rather than for particular billiards. The semiclassical analysis of density of states for quantized chaotic systems results in the Gutzwiller trace formula, which has the great advantage of describing the density of states by using knowledge of nonlinear classical dynamics such as unstable periodic orbits. We then proceed to the framework of thermodynamics of mesoscopic systems like quantum dots, explaining how to incorporate the trace formula into this framework. Since in most mesoscopic systems the number of particles (electrons) is fixed and the thermodynamic limit cannot be taken, we should have recourse to the viewpoint of the canonical ensemble. We show that the Helmholtz free energy in the canonical ensemble is described in terms of more feasible quantities, i.e., thermodynamic functions in the grand canonical ensemble. By studying Chapters 5 through 10, readers will be equipped with enough knowledge to start research work on quantum chaos and quantum transport in quantum dots.

5.1 Semiclassical Quantization of Chaos and Regular Orbits

In contrast to the contents so far, the description from here deals with topical themes, which are rather special and the accompanying explanation might not be self-contained. Readers, if necessary, are recommended to gain supplemental knowledge from references listed at the end of the book. As for the present section, see references (Gutzwiller[5, 8-10], Balian and Bloch[14, 15], Berry and Mount[11], Berry and Tabor[12], Brack and Bhaduri[6], Cvitanović et al.[7]).

Let us consider the stationary state of an electron with mass m_e moving in space of d dimensions. It obeys a quantal one-body Hamiltonian with momentum operator $\boldsymbol{p}\,(=-i\hbar\nabla_{\boldsymbol{q}})$ as

$$\mathcal{H} = \frac{\boldsymbol{p}^2}{2m_e} + V(\boldsymbol{q}). \tag{5.1}$$

(As space coordinates, we choose $\boldsymbol{q}$ and $\boldsymbol{r}$ for general Hamiltonian systems and quantum dots (billiards), respectively.) In (5.1), V denotes the confining potential, which in the case of a hard wall takes $V = 0$ away from the wall, and $V = +\infty$ at the wall. V may be a mean-field potential made by other electrons or an impurity potential. In the presence of a magnetic field, one should employ a vector potential $A(q)$, making the replacement $p \to p + \dfrac{e}{c}A(q)$.

The single-electron energy spectrum is available from an energy-dependent one-particle Green function. The product of one-particle Green functions leads to the two-particle Green function that stores various information about quantum transport. The one-particle Green function to describe an electron propagation from a position q' to another one q'' is given, with the use of eigenvalues $\{E_n\}$ and eigenfunctions $\{\Psi_n\}$ in (5.1), as

$$G^{\pm}(q'',q';E) = \sum_n \frac{\Psi_n(q'')\Psi_n^*(q')}{E - E_n \pm i\varepsilon}. \tag{5.2}$$

Here, $\pm$ denotes the **retarded** $(+)$ and **advanced** $(-)$ **Green function**. In the following, for simplicity, we concentrate on the retarded Green fuction only and abbreviate G^+ and $E + i\varepsilon$ as G and E (complex number), respectively.

Let us take the trace of the Green function as

$$\mathrm{Tr}\,G(E)\ \left(\equiv \int dq\, G(q,q;E)\right) = \sum_n \frac{1}{E - E_n}, \tag{5.3}$$

whose poles give rise to eigenvalues. Noting[6]

$$(x + i\varepsilon)^{-1} = \mathcal{P}x^{-1} - i\pi\delta(x),$$

one obtains the density of states $g(E)$ as follows.

$$g(E) \equiv \sum_n \delta(E - E_n) = -\frac{1}{\pi}\mathrm{Im}\,(\mathrm{Tr}G(E))\,, \tag{5.4}$$

where $\mathrm{Im}(a)$ denotes the imaginary part of a.

Since the Green function (5.2) is nothing but the Laplace transform of the propagator for time evolution

$$K(q'',q';t) \equiv \left\langle q'' \left| e^{-\frac{i}{\hbar}\mathcal{H}t} \right| q' \right\rangle, \tag{5.5}$$

the problem of deriving the Green function is reduced to obtaining the propagator $K(q'',q';t)$.

According to Feynman's path-integral method[3], the propagator K is given by the functional integral

$$K(q'',q';t) = \int_{q(0)=q'}^{q(t)=q''} \mathcal{D}[q]\exp\left(\frac{i}{\hbar}W[q]\right), \tag{5.6}$$

where $W[q]$ is the action function (or Hamilton's principal function) defined in terms of the time integral of the Lagrangian $L = \dfrac{1}{2}m_e\dot{q}^2 - V(q)$ as

---

[6] $\mathcal{P}x^{-1}$ is the principal value (defined except at $x = 0$), and $\delta(x)$ is the **delta function** satisfying $\displaystyle\int_{-\infty}^{\infty}\delta(x)f(x)\,dx = f(0)$ for an arbitrary function $f(x)$.

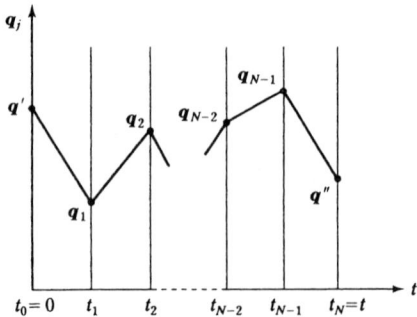

Fig. 5.1   Example of a path in path integrals.

$$W[\boldsymbol{q}] = \int_0^t L(\boldsymbol{q}, \dot{\boldsymbol{q}}) \, dt. \tag{5.7}$$

The meaning of the path integral in (5.6) is as follows. Firstly suppose the time interval t to be divided into N equal segments and introduce the discrete time $t_j = \dfrac{t}{N} j$ (where $j = 0, 1, \cdots, N-1, N$ with $t_0 = 0$, $t_N = t$). Then we approximate the smooth path $\boldsymbol{q}(t)$ with a zigzag path which connects N sectional points $\boldsymbol{q}_j = \boldsymbol{q}(t_j)\,(j = 1, \cdots, N-1)$ (see Fig. 5.1), and calculate $W[\boldsymbol{q}]$ along this zigzag path. Finally, summing the integrand $\exp\left(\dfrac{i}{\hbar} W[\boldsymbol{q}]\right)$ over all possible zigzag paths and taking the limit $N \to \infty$, one can obtain the product of integrals over each sectional point:

$$K(\boldsymbol{q}'', \boldsymbol{q}'; t) = \lim_{N \to \infty} \left(\frac{m_e N}{2\pi i \hbar t}\right)^{d\frac{N}{2}} \int_{-\infty}^{\infty} \cdots \int_{-\infty}^{\infty} \prod_{j=1}^{N-1} d\boldsymbol{q}_j$$

$$\times \exp\left\{\frac{i}{\hbar}\frac{t}{N}\sum_{j=1}^{N}\left[\frac{m_e}{2}\left(\frac{\boldsymbol{q}_j - \boldsymbol{q}_{j-1}}{t/N}\right)^2 - V(\boldsymbol{q}_j)\right]\right\}.$$

$$\tag{5.8}$$

This kind of scheme embodied in (5.6) or (5.8) was proposed by Feynman in 1948 as a new framework of quantum mechanics to be comparable with the Schrödinger or Heisenberg–Dirac formalism.

From the expression in (5.8), we should emphasize that (5.6) implies summation over an infinite number of paths. As already envisaged in Chapter 2, however, by taking the **semiclassical limit** ($\hbar \to 0$), the integrand $\exp\left(\dfrac{i}{\hbar} W[\boldsymbol{q}]\right)$ in (5.6) shows a violent oscillation against a small change in $W[\boldsymbol{q}]$. This phenomenon causes a self-cancellation in the resultant value of the integration (5.6). Eventually, among all possible paths, only those that satisfy the saddle point condition

$$\delta W[\boldsymbol{q}] = 0 \tag{5.9}$$

contribute to the integral. In the stationary-phase approximation, i.e., by incorporating Gaussian fluctuations around the saddle-point solutions, we obtain the well-approximated formula

$$K_{\mathrm{scl}}(\boldsymbol{q}'', \boldsymbol{q}'; t) = \frac{1}{(2\pi i\hbar)^{d/2}} \sum_j \sqrt{\mathcal{J}_j} \exp\left(\frac{i}{\hbar} W_j - \frac{\pi}{2}\sigma_j i\right). \tag{5.10}$$

Here j implies each saddle-point solution with $\boldsymbol{q}(0) = \boldsymbol{q}'$ as an origin and $\boldsymbol{q}(t) = \boldsymbol{q}''$ as a terminal, which is nothing but a classical orbit satisfying Hamilton's least-action principle (5.9), i.e., Hamilton's equation of motion. (5.10) is called the **semiclassical propagator**. Generally there are many orbits (saddle-point solutions). W_j is the action for the classical orbit (path) j between $\boldsymbol{q}'$ and $\boldsymbol{q}''$. $\mathcal{J}_j$ and σ_j are calculated from each orbit j: $\mathcal{J}_j = \left| \det\left(-\dfrac{\partial^2 W_j}{\partial \boldsymbol{q}'' \partial \boldsymbol{q}'} \right) \right|$ is the inverse of the Jacobian arising from quantum fluctuations around j, and σ_j is the number of **conjugate points**[7] of $\mathcal{J}_j$ along j.

As already noted, the Laplace transformation of K_{scl} yields the semiclassical (retarded) Green function

$$G_{\mathrm{scl}}(\boldsymbol{q}'', \boldsymbol{q}'; E) = (i\hbar)^{-1} \int_0^\infty dt K_{\mathrm{scl}}(\boldsymbol{q}'', \boldsymbol{q}'; t) e^{\frac{i}{\hbar} Et}. \tag{5.11}$$

By an additional application of the stationary-phase approximation with the saddle point t_c in the time integration in (5.11), we have

$$G_{\mathrm{scl}}(\boldsymbol{q}'', \boldsymbol{q}'; E)$$
$$= \frac{2\pi}{(2\pi i\hbar)^{(d+1)/2}} \sum_j \sqrt{|\widehat{\mathcal{J}}_j|} \exp\left(\frac{i}{\hbar} S_j(\boldsymbol{q}'', \boldsymbol{q}'; E) - \frac{\pi}{2}\sigma_j i\right). \tag{5.12}$$

In (5.12) the pre-exponential factor is given by

$$\widehat{\mathcal{J}}_j = \widehat{\mathcal{J}}_j(\boldsymbol{q}'', \boldsymbol{q}'; E) = \left(\frac{\partial^2 W_j}{\partial t^2}\right)^{-1} \cdot \det\left(-\frac{\partial^2 W_j}{\partial \boldsymbol{q}'' \partial \boldsymbol{q}'}\right)\Bigg|_{t=t_c}, \tag{5.13}$$

and $S_j(\boldsymbol{q}'', \boldsymbol{q}'; E)$ is a **reduced action** derived from the action function W_j via Legendre transformation as

$$S_j(\boldsymbol{q}'', \boldsymbol{q}'; E) = W_j(\boldsymbol{q}'', \boldsymbol{q}'; t_c) + Et_c$$
$$= \int_{\boldsymbol{q}'}^{\boldsymbol{q}''} \boldsymbol{p}(q, E) \cdot d\boldsymbol{q}. \tag{5.14}$$

Thanks to S, (5.13) can be simplified in the following way. Noting the standard relations in classical mechanics

---

[7] Besides a given original trajectory j, consider new ones with their initial momenta different from the original one by small increments. The fan of these trajectories closes in one or several points, which are called *conjugate points*.

$$\frac{\partial W}{\partial t} = -E, \qquad \frac{\partial S}{\partial E} = t,$$

one finds

$$\frac{\partial^2 W}{\partial t^2} = -\frac{\partial E}{\partial t} = -\left(\frac{\partial^2 S}{\partial E^2}\right)^{-1}.$$

Substituting these formulas into (5.13), we have the renewed expression

$$\widehat{\mathcal{J}}_j(\boldsymbol{q}'', \boldsymbol{q}'; E) = (-1)^d \left(\frac{\partial^2 S_j}{\partial E^2}\right) \det\left(-\frac{\partial^2 W_j}{\partial \boldsymbol{q}'' \partial \boldsymbol{q}'}\right)\Bigg|_{t=t_c}$$

$$= \begin{vmatrix} \dfrac{\partial^2 S_j}{\partial \boldsymbol{q}' \partial \boldsymbol{q}''} & \dfrac{\partial^2 S_j}{\partial \boldsymbol{q}' \partial E} \\[2ex] \dfrac{\partial^2 S_j}{\partial E \partial \boldsymbol{q}''} & \dfrac{\partial^2 S_j}{\partial E^2} \end{vmatrix}, \tag{5.15}$$

where the last form in (5.15) is the determinant of the matrix in $(d+1) \times (d+1)$
dimensions.

Taking the trace of $G_{\mathrm{scl}}(\boldsymbol{q}'', \boldsymbol{q}'; E)$ in (5.12) together with (5.15), we reach the
convenient expression

$$\mathrm{Tr}\, G_{\mathrm{scl}}(E) \left(\equiv \int d\boldsymbol{q} G_{\mathrm{scl}}(\boldsymbol{q}, \boldsymbol{q}; E)\right)$$

$$= \frac{2\pi}{(2\pi i\hbar)^{(d+1)/2}} \sum_j \int d\boldsymbol{q} \sqrt{|\widehat{\mathcal{J}}_j|} \exp\left\{\frac{i}{\hbar} S_j(\boldsymbol{q}, \boldsymbol{q}; E) - \frac{\pi}{2}\sigma_j i\right\}. \tag{5.16}$$

Equations (5.3) and (5.4) with use of (5.16) lead to eigenvalues and the density of
states. While (5.16) holds for both classically integrable and nonintegrable systems,
the following concrete calculation is quite simple in the case of integrable systems, so
we shall manage the trace formula firstly in this case and then proceed to the case of
chaotic systems.

5.1.1 Berry–Tabor's Trace Formula

The system in which the number of independent constants of motion (conserved quantities) accords with the degrees of freedom d is called a **completely integrable system**. In this kind of systems, the phase space is occupied with regular tori like in Fig.
5.2(a), and the actions or adiabatic invariants

$$J_k = \frac{1}{2\pi} \oint_{\Gamma_k} \boldsymbol{p} \cdot d\boldsymbol{q} \qquad (k = 1, 2, \cdots, d) \tag{5.17}$$

corresponding to d **irreducible closed paths** Γ_k play an essential role. Each periodic
orbit winding around a torus proves to be topologically equivalent to a suitable chain
of irreducible closed orbits $\sum_{k=1}^{d} l_k \Gamma_k$ (Fig. 5.2(a)), and the corresponding reduced action
can be written as the sum of integer (l_k) multiples of the irreducible actions J_k.

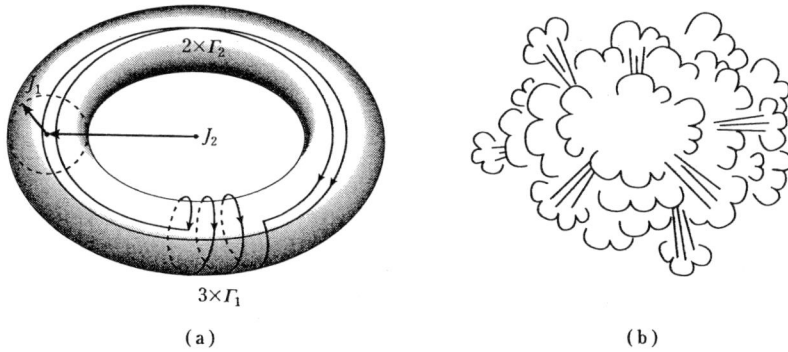

Fig. 5.2 (a) Torus with periodic orbit represented by a chain of irreducible closed paths. In this example, $3\Gamma_1 + 2\Gamma_2$. J_1, J_2 are actions. (b) Collapse of torus.

Let us consider a 2-d system. The canonicl transformation (5.16) from dynamical variables $\boldsymbol{p}, \boldsymbol{q}$ to action-angle variables J_j, φ_j $(j = 1, 2)$ results in a new Hamiltonian $\mathcal{H} = \mathcal{H}(J_1, J_2)$. In the new phase space, the angular frequency is defined as $\omega_j = \dfrac{\partial \mathcal{H}}{\partial J_j}$. Prescribing $\alpha = \dfrac{\omega_1}{\omega_2}$ as the **winding number**, we say the **torus is resonant** if α is rational.

For a given torus, suppose there is a family of periodic orbits characterized by parameters

$$\boldsymbol{L} = (L_1, L_2) = (lu_1, lu_2)$$

to wind around its surface, where mutually prime numbers (u_1, u_2) represent each **primitive periodic orbit** and l denotes its repetition. The winding number for the orbit $\boldsymbol{L}$ can here be written as $\alpha = \dfrac{u_1}{u_2}$. Berry and Tabor[12] showed that only periodic orbits belonging to resonant tori can contribute to the integral in (5.16).

To mention their outcome, the trace of the Green function $\mathrm{Tr}\, G_{\mathrm{scl}}$ consists of the sum of the **mean part** (contributions due to zero-period orbits), $\mathrm{Tr}\, G_0$, and the **oscillating part** (contributions due to periodic orbits), $\delta \mathrm{Tr}\, G_{\mathrm{scl}}$, with the latter given by

$$\delta \mathrm{Tr}\, G_{\mathrm{scl}}(E) = \sum_{\boldsymbol{L} \neq (0,0)} \frac{-i\tau_{\boldsymbol{L}}}{\hbar^{3/2} L_2^{3/2} \left| F_E^{(\prime\prime)}(J_1^{\boldsymbol{L}}) \right|^{1/2}} \exp\left[i\left(\frac{S_{\boldsymbol{L}}}{\hbar} - \mu_{\boldsymbol{L}} \frac{\pi}{2} + \gamma \frac{\pi}{4} \right) \right].$$

(5.18)

In (5.18), μ is the Maslov index generalized so as to include the number of conjugate points σ (see the next subsection), and S and τ are, respectively, the action and period (for a family of periodic orbits $\boldsymbol{L}$ with energy E) with their explicit forms given by

$$S_{\boldsymbol{L}} = 2\pi l(u_1 J_1 + u_2 J_2), \qquad \tau_{\boldsymbol{L}} = \frac{2\pi l u_1}{\omega_1}.$$

The structure of the torus is determined via $J_2 = F_E(J_1)$, which is obtained by rewriting $\mathcal{H}(J_1, J_2) = E$. For a given orbit $\boldsymbol{L}$, the winding number α and phase factor γ in (5.18) are derived by the first and second derivatives of F_E, respectively, as

$$\alpha = -F_E^{(\prime)}(J_1^{\boldsymbol{L}}), \qquad \gamma = \operatorname{sgn}\left(F_E^{(\prime\prime)}(J_1^{\boldsymbol{L}})\right).$$

Consequently, the oscillating part of the density of states is given by

$$\delta g(E) = -\pi^{-1}\operatorname{Im}\left(\delta\operatorname{Tr}G_{\mathrm{scl}}(E)\right).$$

(5.18) is the **Berry–Tabor trace formula** for integrable systems[12] .

Originally, this formula was reached by applying (i) the Poisson summation formula and (ii) saddle-point integration to the very definition of density of states for a system to which the Einstein–Brillouin–Keller (EBK) quantization rule is applicable. Namely it assumes the presence of tori and exploits the advantage of the action as a constant of motion. A crucial point of (5.18) lies in that a quantum-mechanical quantity (its left-hand side) is described by using only knowledge of periodic orbits (the right-hand side).

5.1.2 Gutzwiller's Trace Formula

For chaotic systems, on the other hand, Gutzwiller pursued a corresponding trace formula and semiclassical quantization condition[5, 8-10]. In chaotic systems, the transformation from $\boldsymbol{p}, \boldsymbol{q}$ to action-angle variables is no longer meaningful because of the collapse of tori (see Fig. 5.2(b)). In this case, we further carry out the saddle-point approximation in $\boldsymbol{q}$ integration in (5.16), arriving at

$$\frac{\partial S(\boldsymbol{q}, \boldsymbol{q})}{\partial \boldsymbol{q}} \equiv \frac{\partial S(\boldsymbol{q}'', \boldsymbol{q}')}{\partial \boldsymbol{q}''} + \left.\frac{\partial S(\boldsymbol{q}'', \boldsymbol{q}')}{\partial \boldsymbol{q}'}\right|_{\boldsymbol{q}''=\boldsymbol{q}'=\boldsymbol{q}} = \boldsymbol{p}'' - \boldsymbol{p}' = 0,$$

$$(5.19)$$

where $\boldsymbol{p}'$ and $\boldsymbol{p}''$ are momenta at the origin and terminal, respectively. Combining the condition for momenta $\boldsymbol{p}'' = \boldsymbol{p}'$, (5.19), together with the tracing condition, $\boldsymbol{q}'' = \boldsymbol{q}' = \boldsymbol{q}$, we come to understand that only periodic orbits (hereafter to be often abbreviated as PO) contribute to $\operatorname{Tr}G_{\mathrm{scl}}(E)$. (Paradoxically, there is no contribution due to chaotic orbits.) In fact, in **fully chaotic systems without bifurcations** one can find isolated unstable periodic orbits with positive Lyapunov exponents. The phase space of a fully chaotic system is occupied by a **chaotic sea**, i.e., filled with chaotic orbits. Despite this fact, one can find **isolated unstable periodic orbits in the chaotic sea**, which are of zero measure in the chaotic sea but infinite in number. Thanks to (5.19), therefore, (5.16) can now be expressed as a sum of contributions arising from all unstable periodic orbits.

To attain the Gutzwiller formula (5.29), we shall first rewrite the pre-exponential factor of (5.16) (Jacobian (5.15)) by using global characterisic quantities associated with each periodic orbit. (Readers wishing to quickly reach the goal (5.29) may skip the description from here up to the first paragragh just after (5.28).)

For each periodic orbit (PO), let us introduce local coordinates with components $q_{//}$ and $q_\perp$, respectively, parallel and perpendicular to PO as

$$q = (q_{//}, q_{\perp 1}, q_{\perp 2}, \cdots, q_{\perp(d-1)}) = (q_{//}, q_\perp). \tag{5.20}$$

The law of energy conservation in the form of the Hamilton–Jacobi equation

$$\mathcal{H}(q, p) = \mathcal{H}\left(q, \frac{\partial S}{\partial q}\right) = E$$

leads to the convenient identity

$$\left| \frac{\partial^2 S}{\partial q' \partial q''} \right| = 0, \tag{5.21}$$

which greatly simplifies (5.15) as follows:

$$\begin{aligned}
\widehat{\mathcal{J}}_{po} &= -\frac{\partial^2 S_{po}}{\partial E \partial q''_{//}} \frac{\partial^2 S_{po}}{\partial E \partial q'_{//}} \left| \frac{\partial^2 S_{po}}{\partial q'_\perp \partial q''_\perp} \right| \\
&= (-1)^d \frac{1}{\dot{q}''_{//} \dot{q}'_{//}} \det\left(\frac{\partial p'_\perp}{\partial q''_\perp} \right).
\end{aligned} \tag{5.22}$$

Besides (5.22), there will appear an extra pre-exponential factor in (5.16). Around the saddle point provided by (5.19), we expand the reduced action up to second order in $q_\perp$, having

$$S(q'', q', E)|_{q''=q'=q} = S_{po}(E) + \frac{1}{2} \sum_{i,j=1}^{d-1} R_{ij}(q_{//}) q_{\perp i} q_{\perp j} \tag{5.23}$$

with $S_{po}(E)$ an action for the PO,

$$S_{po}(E) = \oint_{po} p \cdot dq, \tag{5.24}$$

and $\{R_{ij}(q_{//})\}$ the second-order mixed derivatives of $S(q'', q', E)$ with respect to $q_\perp$, representing elements of a $(d-1) \times (d-1)$ dimensional matrix, i.e.,

$$R(q_{//}) = \left(\frac{\partial^2 S}{\partial q''_\perp \partial q''_\perp} + \frac{\partial^2 S}{\partial q''_\perp \partial q'_\perp} + \frac{\partial^2 S}{\partial q'_\perp \partial q''_\perp} + \frac{\partial^2 S}{\partial q'_\perp \partial q'_\perp} \right)_{q''_\perp = q'_\perp = 0}. \tag{5.25}$$

Substitution of (5.23) into (5.16) followed by Gaussian integration over $q_\perp$ completes the saddle-point integration in (5.16). The contribution of each PO to the oscillating part of the trace is eventually given by

$$\delta \mathrm{Tr}\, G_{scl}(E) \propto \exp\left\{ i \left(\frac{S_{po}}{\hbar} - \nu_{po} \right) \right\}$$

$$\times \int \left|\widehat{\mathcal{J}}_{po}(q_{//})\right|^{1/2} \left|\det R(q_{//})\right|^{-1/2} dq_{//}. \tag{5.26}$$

Here, the phase factor ν_{po} denotes $\dfrac{\pi}{2} \times \mu_{po}$ with μ_{po} the Maslov index which is the sum of the number of conjugate points σ_{po} and that of the negative eigenvalues of (5.25).

We proceed to rewrite the integral in (5.26) in terms of the **linearized Poincaré map** (the **monodromy matrix**) vital to the study of the stability of the PO. If one uses in (5.22) and (5.25) the relations

$$\frac{\partial S}{\partial q''} = p'', \qquad \frac{\partial S}{\partial q'} = -p',$$

the inverse of the integrand in (5.26) is reduced to

$$\left|\widehat{\mathcal{J}}_{po}(q_{//})\right|^{-1/2} \left|\det R(q_{//})\right|^{1/2}$$

$$= |\dot{q}_{//}| \times \left| \frac{\partial(q''_{\perp} - q'_{\perp}, p''_{\perp} - p'_{\perp})}{\partial(q'_{\perp}, p'_{\perp})} \right|^{1/2}_{q''_{\perp} = q'_{\perp} = 0}. \tag{5.27}$$

The right-hand side of (5.27), aside from a variable $\dot{q}_{//}$, is an invariant proper to each PO. Guided by the supplemental $2(d-1)$ dimensional vector $\boldsymbol{\xi}_{\perp} = (q_{\perp}, p_{\perp})$ defined in phase space, this invariant turns out to be related to the $2(d-1) \times 2(d-1)$ dimensional **stability matrix** $\tilde{M}_{po}$

$$\left| \frac{\partial(q''_{\perp} - q'_{\perp}, p''_{\perp} - p'_{\perp})}{\partial(q'_{\perp}, p'_{\perp})} \right|_{q''_{\perp} = q'_{\perp} = 0} = \left| \frac{\partial(\boldsymbol{\xi}''_{\perp} - \boldsymbol{\xi}'_{\perp})}{\partial \boldsymbol{\xi}'_{\perp}} \right|_{q''_{\perp} = q'_{\perp} = 0}$$

$$= \left| \det(\tilde{M}_{po} - I) \right|. \tag{5.28}$$

$\tilde{M}_{po}$ is the nontrivial part of the monodromy matrix. It characterizes the transverse instability of the PO, describing how a transversal deviation of the PO at the origin ($q' = 0$) would be amplified at the end ($q'' = 0$). Noting also the additional invariant

$$\int \frac{1}{\dot{q}_{//}} dq_{//} = \int_{\text{primitive PO}} dt = T_p,$$

the integral in (5.26) proves to be expressed in terms of global invariant quantities of the PO. T_p above means the period of the primitive PO that goes round it once.

Summing up the contributions (5.26) from all PO's, the calculation of (5.16) is completed. Supposing each PO to be an integer (l_p) multiple of a primitive PO, p, with the action and period defined, respectively, by

$$S_p(E) \left(= \oint_p \boldsymbol{p} \cdot d\boldsymbol{q}\right) \quad \text{and} \quad T_p(E) \left(= \frac{\partial S_p}{\partial E}\right),$$

at last we reach our goal: the oscillating part of the trace of the Green function, in the case of nonintegrable and chaotic systems, is expressed as

$$\delta \operatorname{Tr} G_{\mathrm{scl}}(E) \equiv \operatorname{Tr} G_{\mathrm{scl}}(E) - \operatorname{Tr} G_0(E)$$
$$= \frac{1}{i\hbar} \sum_p T_p \sum_{l_p=1}^{\infty} \left(\det(\tilde{M}_p^{l_p} - I)\right)^{-1/2} \exp\left\{il_p\left(\frac{1}{\hbar}S_p - \nu_p\right)\right\}$$
$$+ O\left(\hbar^0\right). \tag{5.29}$$

$O(\hbar^0)$ signifies the magnitude of the corrections being at most of order $\hbar^0$. (5.29) is the so-called **Gutzwiller trace formula**. It should be noted that since we have assumed nothing other than the isolated PO's (no degeneracy of saddle points), the Gutzwiller formula should also hold for completely integrable systems with neither continuous symmetry nor orbit bifurcations. Such an example is the Berry–Tabor trace formula in (5.18). In (5.29), $\operatorname{Tr} G_0(E)$ is the mean part of the trace,

$$\operatorname{Tr} G_0(E) = \frac{\pi}{i(2\pi\hbar)^d} \iint d\boldsymbol{p} d\boldsymbol{q} \delta(H(\boldsymbol{q},\boldsymbol{p}) - E) + O(\hbar^{-d+1}) \tag{5.30}$$

coming from the zero-length orbit. $\operatorname{Tr} G_0(E)$ is proportional to the mean density, i.e., the volume of the energy shell in phase space divided by the volume $(2\pi\hbar)^d$ of the Planck cell (the d-dimensional supercube with linear dimension of $2\pi\hbar$). Correspondingly, while the oscillating part of density of states is derived as

$$\delta g(E) = -\pi^{-1}\operatorname{Im}\left(\delta \operatorname{Tr} G_{\mathrm{scl}}(E)\right), \tag{5.31}$$

its mean part is

$$\bar{g}(E) = -\pi^{-1}\operatorname{Im}\left(\operatorname{Tr} G_0(E)\right). \tag{5.32}$$

As already noted above, the stability matrix (linearized Poincaré map), $\tilde{M}_p$, captures the time evolution of the variation $\delta \boldsymbol{q}_\perp^{(n)}$ perpendicular to each orbit p:

$$\delta \boldsymbol{q}_\perp^{(n+1)} = \tilde{M}_p \delta \boldsymbol{q}_\perp^{(n)}. \tag{5.33}$$

The eigenvalues Λ_p of $\tilde{M}_p$ depend on the types of PO (or fixed points in the case of a map), and take values $\Lambda_p = \exp(\pm\lambda_p)$ and $\exp(i\lambda_p)$ for unstable and stable POs, respectively. For an unstable PO, in particular, we have

$$\det(M_p^l - I) = 4\sinh^2\left(\frac{l\lambda_p}{2}\right). \tag{5.34}$$

In this and the following chapters, we shall apply the Gutzwiller trace formula to various topics of mesoscopic physics. One should, however, keep in mind a big problem involved in this formula. Recall the KS entropy h_{KS} characterizing the degree of randomness of a dynamical system, which, as shown in Chapter 3, is equal to the Lyapunov exponent in the case of closed billiard systems. With the use of h_{KS}, the total number of POs with period T ($\gg$ typical time of a given system) is described as

$$N(T) \sim \frac{1}{T} \exp\left(h_{KS}T\right),$$

whereas the amplitude (pre-exponential factor) of the term in (5.29) for a PO with period T is approximately

$$A(T) \sim T \exp\left(-\frac{h_{KS}T}{2}\right).$$

Combining these two facts, the contributions in (5.29) due to a PO with period less than T amount to

$$A(T) \times N(T) \sim \exp\left(\frac{h_{KS}T}{2}\right),$$

showing an exponential divergence with T so long as one adds terms in (5.29) simply in order of increasing period of PO. This exponential proliferation is a big problem in semiclassical quantization of chaotic systems. One of the most important subjects of quantum chaos is a search to converge the periodic orbits sum, by applying, e.g., a method of re-summation, and to obtain a reliable semiclassical density of states and eigenvalues.

In practical problems encountered in mesoscopic physics, however, we deal with systems at finite temperatures, incorporate the trace formula into the framework of thermodynamics, and eventually can avoid the mathematical problem of exponential proliferation. This will be shown vividly in the next section.

Before closing this section, we should comment that the formula (5.29) itself breaks down in the cases that the amplitude of the pre-exponential factor in (5.29) diverges. This divergence occurs when saddle points (periodic orbits) become degenerate. Examples are (i) systems that have continuous symmetry and (ii) those showing orbit bifurcations. The case (i) can be resolved by moving first to general phase-space variables and then carrying out the trace integral exactly over the measure of the group that characterizes the continuous symmetry (Creagh and Littlejohn[54,55]). The case (ii) is solved by going to higher orders in the saddle-point expansion (uniform approximation) and then performing the diffraction-catastrophe integrals (Sieber[56,57], Schomerus and Sieber[58]).

We explain the case (i) in more detail and indicate a general way of deriving the improved trace formula for systems with continuous symmetry. Let a system with d degrees of freedom have a continuous symmetry characterized by f independent parameters (variables). For example, the systems with $U(1)$ symmetry and with the spherical $SO(3)$ symmetry have a single ($f = 1$) polar angle and three ($f = 3$) Euler angles, respectively. PO's of these systems appear as continuous degenerate families with each family specified by f group parameters. Noting that the time

t – the variable for the continuous time translation – also joins the above group parameters, the periodic orbit families cover an $(f + 1)$-dimensional hypersurface in phase space. (The variable t stands for the space coordinate along the trajectory of PO.) The sum over PO's in the trace formula now becomes a sum over discrete orbit families labeled by Γ, each containing the $(f + 1)$-dimensional integral over the time t and f group parameters. To derive the density of states $g(E)$ from the position-coordinate representation of the Green function, $G(q', q; E)$, we must take the trace over q space in d dimensions. It should be noted that the method of integration depends on the value of $f + 1$. If $f + 1 \leq d$, we first carry out the exact $f + 1$ group integrations and then the remaining $d - (f + 1)$ spatial integrations in the stationary-phase approximation. On the other hand, if $f + 1 > d$, one must be careful not to integrate unphysically beyond d-dimensional space. Creagh and Littlejohn employed a mixed representation of the Green function, $G(p', q; E)$, with the final momentum (p') and the initial coordinate q. The density of states is then given by the trace over the phase space in $2d$ dimensions: it is now possible to integrate over $f + 1$ variables exactly with the remaining $2d - (f + 1)$ integrations (transverse to the orbit families) in the stationary-phase approximations.

The improved trace formula for systems with f dimensional degeneracy is[54, 55]

$$\delta g(E) = -\frac{1}{\pi} \mathrm{Im} \left(\frac{1}{i\hbar} \frac{1}{(2\pi i\hbar)^{f/2}} \sum_\Gamma \int_\Gamma dt d\mu(f) |\mathcal{M}_\Gamma|^{-1/2} \exp \left[i \left(\tfrac{1}{\hbar} S_\Gamma - \tfrac{\pi}{2} \mu_\Gamma \right) \right] \right).$$

$$(5.35)$$

The discrete sum is not for isolated PO's but for distinct families Γ of PO's; the action S_Γ and Maslov index μ_Γ are also for Γ. Here Γ means both each primitive periodic orbit family and its repetitions. The continuous summation within each family is treated by the measure $dt d\mu(f)$. The factor $\mathcal{M}_\Gamma$ represents the stability of one typical orbit of each Γ. The formula (5.35) is the extension of the Gutzwiller trace formula. In fact, in the limit of no continuous symmetry ($f = 0$), $\int_\Gamma dt d\mu(f) \to \int_\Gamma dt = T_p$ and $\mathcal{M}_\Gamma \to \det(\tilde{M}_p^{l_p} - I)$, which recovers the oscillating part (5.31) with (5.29). The amplitude of the density of states in (5.35) has a factor of $\hbar^{-(\frac{f}{2}+1)}$, growing with increasing degree of symmetry f.

We shall come back to the concept of continuous symmetry and that of orbit bifurcations in the context of quantum dots in Chapters 6 and 9, respectively.

5.2   Thermodynamics of Mesoscopic Systems

In the thermodynamics of mesoscopic systems like quantum dots and antidots, the number of particles (electrons) N, though $N \gg 1$, is finite and fixed. Hence one cannot assume the thermodynamic limit ($N \to \infty$, V(volume)$\to \infty$, $\mathcal{A}$(area)$\to \infty$). In these circumstances, one must choose the standpoint of the canonical ensemble. On the contrary, the grand canonical ensemble with a fixed chemical potential assumes no constraint on N, and has the advantage of allowing much simpler calculations of thermodynamic quantities. In what follows, we shall first derive the thermodynamic functions within the framework of the grand canonical ensemble, and then derive the free energy within the framework of the canonical ensemble[6, 27, 60, 61]. The description below follows work by Richter, Ullmo and Jalabert[60].

5.2.1   Grand Canonical Ensemble

As we found in the semiclassical trace formula, the density of states g is the sum of a mean part $\bar{g}$ and an oscillating part δg. While $\bar{g}$ is determined only by universal quantities like the dimensionality and volume (area) of systems, δg depends largely on the nature of periodic orbits, i.e., the system's specific features such as billiard shapes and nonlinear dynamics. This means that a variety of features of both quantum transport and thermodynamic quantities are affected only by δg. We shall hereafter demonstrate how δg is incorporated into the framework of thermodynamics. As we saw in the previous section, δg is expressed as

$$\delta g(E) = \sum_{po} \delta g_{po}(E)$$

$$= \sum_{po} A_{po}(E) \cos\left(\frac{S_{po}(E)}{\hbar} - \nu_{po}\right), \tag{5.36}$$

irrespective of whether the system is integrable or chaotic. In (5.36) the PO is prescribed to denote both the primitive periodic orbit and its multiples. If one wishes to take into consideration the spin degree of freedom also, one may simply multiply the density of states by the spin factor g_s ($= 2$). By integrating (5.36) over the energy, we have an oscillating part of the number of states (accumulated density of states) for Fermi systems at absolute zero temperature:

$$\delta n(E) = \int_0^E dE' \delta g(E')$$

$$= \sum_{po} \int_0^E dE' A_{po}(E') \cos\left(\frac{S_{po}(E')}{\hbar} - \nu_{po}\right). \tag{5.37}$$

Throughout this book, except when stated otherwise, the semiclassical approximation is used where only terms of leading order in the Planck constant $\hbar$ are employed. Then, in each of the integrations that follow, one may ignore the variation of amplitude as compared with that of the phase part ("cos" factor). Thanks to a variable transformation (from E to S_{po}) based on $\dfrac{dS_{po}}{dE} = \tau_{po}(E)$, we can carry out the integration in (5.37), to obtain

$$\delta n(E) = \sum_{po} \frac{\hbar}{\tau_{po}(E)} A_{po}(E) \sin\left(\frac{S_{po}(E)}{\hbar} - \nu_{po}\right)$$

$$= \sum_{po} \frac{\hbar}{\tau_{po}(E)} \delta g^{\sharp}_{po}(E), \tag{5.38}$$

where $\sharp$ denotes a phase shift by $-\dfrac{\pi}{2}$.

Following the same procedure as above, the oscillating part of the thermodynamic potential for Fermi systems at absolute zero is given by

$$\delta\omega(E) = -\int_0^E dE' \delta n(E')$$

$$= \sum_{po} \left(\frac{\hbar}{\tau_{po}(E)}\right)^2 \delta g_{po}(E). \tag{5.39}$$

The next step is to calculate thermodynamic functions for systems at **finite temperatures** as yet within the grand canonical ensemble. Let us introduce the inverse temperature $\beta = \dfrac{1}{k_{\mathrm{B}}T}$, the chemical potential (the energy required to add a particle to the system) μ, and the **density of states at finite temperatures** $D(\mu)$. $D(\mu)$ can also be written as a sum of the mean and oscillating parts, $D(\mu) = \overline{D}(\mu) + \delta D(\mu)$. We shall confine ourselves to a realistic temperature regime ($\Delta \ll k_{\mathrm{B}}T \ll \mu$) which is higher than the mean level separation (Δ) but less than μ. Using the **Fermi–Dirac distribution function**

$$f(E - \mu) = [1 + \exp(\beta(E - \mu))]^{-1} \tag{5.40}$$

together with its derivative $f'(E - \mu)$, $\delta D(\mu)$ is defined as

$$\delta D(\mu) = \frac{d}{d\mu} \int_0^\infty dE \delta g(E) f(E - \mu)$$

$$= -\int_0^\infty dE \delta g(E) f'(E - \mu). \tag{5.41}$$

The last integral in (5.41) can be evaluated using a complex-integral method (see Fig. 5.3). In the complex E plane, $f'(E - \mu)$ has period $\dfrac{2\pi i}{\beta}$ along the $\Im$ axis, and there exist second-order poles at $E_n = \mu + \dfrac{(2n+1)\pi i}{\beta}$ with n an arbitrary integer. Let us extend the integration path in the last integral of (5.41) to a (counter-clockwise) rectangular-shaped contour of integration as

$$\int_0^\infty \longrightarrow \int_0^\infty + \int_\infty^{\infty + \frac{2\pi i}{\beta}} + \int_{\infty + \frac{2\pi i}{\beta}}^{0 + \frac{2\pi i}{\beta}} + \int_{0 + \frac{2\pi i}{\beta}}^0, \tag{5.42}$$

inside which only a single pole exists. Further, in the temperature range $k_{\mathrm{B}}T \ll \mu$, one observes $f'(E - \mu)$ to be vanishing except in the narrow strip along the $\Im$ axis

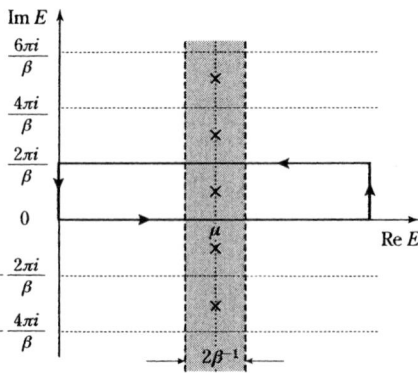

Fig. 5.3   Complex energy plane and contour of integration needed to calculate the last integral in (5.41). × denotes a pole of $f'(E - \mu)$.

$(\mu - \beta^{-1} < \text{Re } E < \mu + \beta^{-1})$, and thereby one can suppress the contributions from the second and fourth integrals in (5.42). Noting all these facts, the contour integration for each periodic orbit that contributes to δD is written as follows:

$$\int_0^\infty dE A_{po}(E) \exp\left[\frac{i}{\hbar} S_{po}(E)\right] f'(E - \mu)$$

$$+ \int_{\infty + \frac{2\pi i}{\beta}}^{0 + \frac{2\pi i}{\beta}} dE A_{po}(E) \exp\left[\frac{i}{\hbar} S_{po}(E)\right] f'(E - \mu)$$

$$= \frac{i}{\hbar} \tau(E_0) \frac{2\pi i}{\beta} A_{po}(E_0) \exp\left[\frac{i}{\hbar} S_{po}(E_0)\right], \qquad (5.43)$$

where the right-hand side represents a residue coming from the pole $\left(E_0 = \mu + \dfrac{\pi i}{\beta}\right)$ inside the contour of integration. In the calculation of (5.43), the change of amplitude factor against energy has been assumed negligible as compared with that of the phase factor.

In the following, we shall take into consideration only the leading term in $\hbar$ and β^{-1}. On the left-hand side of (5.43), the second integration can be reduced to one along the $\Re$ axis owing to the periodicity of $f'(E - \mu)$ with period $\dfrac{2\pi i}{\beta}$, and can be merged with the first one as

$$\text{l.h.s. of (5.43)} = \int_0^\infty dE \left\{ A_{po}(E) \exp\left[\frac{i}{\hbar} S_{po}(E)\right] \right.$$

$$\left. - A_{po}\left(E + \frac{2\pi i}{\beta}\right) \exp\left[\frac{i}{\hbar} S_{po}\left(E + \frac{2\pi i}{\beta}\right)\right] \right\} f'(E - \mu).$$

$$(5.44)$$

Using in (5.43) and (5.44) the approximations

$$S_{po}\left(E+\frac{2\pi i}{\beta}\right) \sim S_{po}(E) + \tau(\mu)\frac{2\pi i}{\beta} \; ; \quad S_{po}(E_0) \sim S_{po}(\mu) + \tau(\mu)\frac{\pi i}{\beta},$$

and ignoring the corresponding correction in the pre-exponential factor, one finds that (5.43) is reduced to

$$\left(1 - \exp\left(-\frac{2\pi\tau(\mu)}{\beta\hbar}\right)\right) \int_0^\infty dE\, A_{po}(E) \exp\left[\frac{i}{\hbar}S_{po}(E)\right] f'(E-\mu)$$

$$= -\frac{2\pi\tau(\mu)}{\beta\hbar} A_{po}(\mu) \exp\left(-\frac{\pi\tau(\mu)}{\beta\hbar}\right) \exp\left(\frac{i}{\hbar}S_{po}(\mu)\right),$$

(5.45)

which is rewritten in a more elegant form:

$$\int_0^\infty dE\, A_{po}(E) \exp\left[\frac{i}{\hbar}S_{po}(E)\right] f'(E-\mu)$$

$$= -A_{po}(\mu) \exp\left[\frac{i}{\hbar}S_{po}(\mu)\right] \Xi\left(\frac{\tau_{po}}{t_T}\right).$$

(5.46)

Ξ is a temperature-dependent damping factor defined by

$$\Xi\left(\frac{\tau_{po}}{t_T}\right) = \frac{\dfrac{\tau_{po}}{t_T}}{\sinh\left(\dfrac{\tau_{po}}{t_T}\right)}$$

(5.47)

with $t_T = \dfrac{\beta\hbar}{\pi}$ the cutoff time due to the finite temperatue effect. In (5.46), the remaining factor prior to Ξ is nothing but the zero-temperature value of the left-hand side of (5.46).

Collecting contributions from all periodic orbits, we obtain

$$\delta D(\mu) = \sum_{po} \delta g_{po}(\mu) \,\Xi\left(\frac{\tau_{po}}{t_T}\right).$$

(5.48)

The damping factor may also be defined in terms of a length variable. In this case, with the use of a thermal cutoff length $L_T = \dfrac{\hbar v_F \beta}{\pi}$, another kind of definition is possible as

$$\Xi\left(\frac{L_{po}}{L_T}\right) = \frac{\dfrac{L_{po}}{L_T}}{\sinh\left(\dfrac{L_{po}}{L_T}\right)}.$$

(5.49)

Observing the exponential decay $\Xi(x) \sim 2xe^{-x}$ at $x \gg 1$, it is evident that only periodic orbits with period (length) less than $t_T(L_T)$ contribute to thermodynamic functions. This gives rise to the important notion that in our application of the trace formula to mesoscopic systems, the finite-temperature effect suppresses contribution from long orbits, which practically sweeps away the problem of exponential proliferation.

Similarly, the particle number $N(\mu)$ and thermodynamic potential $\Omega(\mu)$ for Fermi systems at finite temperatures have their oscillating part,

$$\delta N(\mu) = -\int_0^\infty dE \delta n(E) f'(E - \mu)$$

$$= \sum_{po} \frac{\hbar}{\tau_{po}(\mu)} \delta g_{po}^\sharp(\mu) \Xi \left(\frac{\tau_{po}}{t_T} \right), \tag{5.50}$$

$$\delta \Omega(\mu) = -\int_0^\infty dE \delta \omega(E) f'(E - \mu)$$

$$= \sum_{po} \left(\frac{\hbar}{\tau_{po}(\mu)} \right)^2 \delta g_{po}(\mu) \Xi \left(\frac{\tau_{po}}{t_T} \right). \tag{5.51}$$

5.2.2 Canonical Ensemble

In mesoscopic systems like quantum dots, the particle (electron) number is fixed, i.e., "N = constant", with the system's volume V (area $\mathcal{A}$ in 2-d systems) kept finite. This fact means that the condition "N = constant" is not identical to the condition, "chemical potential μ = constant", making the familiar formulas in the thermodynamic limit apparently useless. The thermodynamics for systems with "N = constant" should be constructed on the basis of the canonical ensemble. Here, the core of thermodynamic functions is the Helmholtz free energy F, which is related to the thermodynamic potential Ω in the grand canonical ensemble through the Legendre transformation

$$F(T, B, N) = \mu N + \Omega(T, B, \mu), \tag{5.52}$$

where B typically represents an external field. At finite temperatures, as pointed out in (5.48) and (5.49), only short periodic orbits contribute to the oscillating part of the density of states δD, which guarantees $\delta D \ll \overline{D}$ and makes possible an expansion of F in the small parameter $\dfrac{\delta D}{\overline{D}}$.

To be explicit, let us concentrate on the chemical potential μ and derive a **mesoscopic correction** $\delta\mu$ to the mean value $\overline{\mu}$ which is defined as follows. First, decompose the μ-dependent particle number $N(\mu)$

$$N(\mu) = \int_0^\infty dE g(E) f(E - \mu)$$

formally into mean and oscillating parts as $N = \overline{N} + \delta N$, and fill N (fixed value!) particles in the system characterized by the mean values $\overline{\mu}$ and $\overline{D}$ (see Fig. 5.4). Then

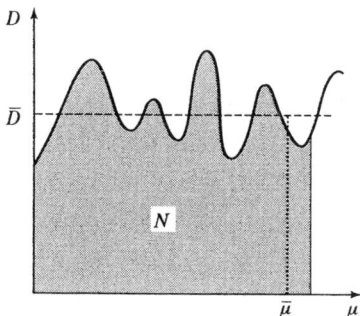

Fig. 5.4 Decomposition of the density of states $D(\mu)$ (solid line) into the mean part $\overline{D}$ (broken line) and the oscillating part δD (fluctuations around the broken line) in the case of 2-d systems. The particle number N (gray area) equals $\overline{D} \times \overline{\mu}$.

we obtain the equation which defines $\overline{\mu}$:

$$N \equiv N(\mu) = \overline{N}(\overline{\mu}). \tag{5.53}$$

$\overline{N}(\overline{\mu})$ implies the value calculable from the equation just above (5.53) with substitutions $g(E) \to \overline{g}(E)$, $\mu \to \overline{\mu}$ into its integrand. On the other hand, the expansion of (5.53) up to linear term in $\delta\mu$ $(\equiv \mu - \overline{\mu})$, together with use of $\dfrac{dN}{d\mu} = D$, gives the mesoscopic correction

$$\delta\mu \sim -\frac{\delta N(\overline{\mu})}{\overline{D}(\overline{\mu})} \qquad (\delta N(\overline{\mu}) \equiv N(\overline{\mu}) - N). \tag{5.54}$$

Everything now gets ready for proceeding to the evaluation of the Helmholtz free energy. Let us expand F to quadratic terms in $\delta\mu$ with the help of the relation $\dfrac{\partial\Omega}{\partial\mu} = -N$, and then we have

$$F(T, B, N) = (\overline{\mu} + \delta\mu)N + \Omega(\overline{\mu}) - N(\overline{\mu})\delta\mu - \frac{1}{2}D(\overline{\mu})\delta\mu^2. \tag{5.55}$$

Using the additional decomposition $\Omega(\overline{\mu}) = \overline{\Omega}(\overline{\mu}) + \delta\Omega(\overline{\mu})$, together with relations (5.53) and (5.54), the expansion of F in $\delta\mu$ in (5.55) is reduced to[60]

$$F(T, B, N) = F^{(0)} + \delta F^{(1)} + \delta F^{(2)} \tag{5.56}$$

with

$$F^{(0)} = \overline{\mu}N + \overline{\Omega}(\overline{\mu}), \qquad \delta F^{(1)} = \delta\Omega(\overline{\mu}), \qquad \delta F^{(2)} = \frac{(\delta N(\overline{\mu}))^2}{2\overline{D}(\overline{\mu})}. \tag{5.57}$$

$\delta F^{(1)}, \delta F^{(2)}$ are the so-called mesoscopic corrections , which are fortunately calculable using the mean and oscillating parts already involved in the grand canonical ensemble.

$F^{(0)} + \delta F^{(1)}$ can give susceptibility and persistent current, etc. in the canonical ensemble, calculated through the grand canonical quantities. At this stage, however, no clear distinction between the canonical and grand canonical ensembles is found, because of the oscillatory and self-cancelling property of $\delta\Omega$. A crucial correction proper to the canonical ensemble is provided by $\delta F^{(2)}$. Physical quantities such as persistent current are derived from $F^{(0)} + \delta F^{(1)}$, which, after being averaged over the Fermi energy or sample size, give an **individual value** for the system. This value can be vanishing, however, because the average $\langle \delta\Omega \rangle \sim 0$ is possible due to the oscillating nature of $\delta\Omega$. By contrast, quantities derived from $\delta F^{(2)}$ have nonvanishing finite values thanks to $\langle (\delta N)^2 \rangle \neq 0$ and are called **ensemble-mean values**.

In this chapter, we found that the semiclassical representation of the Green function makes it possible to capture quantum-mechanical signatures of chaos or regular trajectories. The oscillating part of the density of states δg is represented by global quantities of peridic orbits only, and is dependent largely on the integrability or non-integrability of the underlying nonlinear dynamics. In particular, the Gutzwiller formula for δg, applied to chaotic systems, meets the problem of exponential divergence because of the exponential growth of long periodic orbits. Once combined with the thermodynamics for mesoscopic systems, however, this formula is multiplied with a thermal damping factor that suppresses contributions from long periodic orbits, and the problem of divergence is eventually overcome. We have also presented an essential mesoscopic correction to the Helmholtz free energy, which is proper to the canonical ensemble of mesoscopic systems.

In Chapter 6, we shall apply the formalism described here to thermodynamic properties of closed quantum dots.

6

ORBITAL DIAMAGNETISM AND PERSISTENT CURRENT

The orbital diamagnetism is an essential concept by means of which to understand topical phenomena in modern physics. The criterion for superconductivity is complete diamagnetism (the Meissner effect) and the quantum Hall effect is explained in terms of diamagnetic edge currents. In this chapter we shall investigate such diamagnetism from the novel viewpoint of chaos and nonlinear dynamics in general. We start with introducing the historical background of diamagnetism for an electron gas and then proceed to apply the semiclassical theory of the previous chapter to the calculation of the diamagnetic susceptibility χ. We shall exhibit how the dependence of χ on the Fermi energy and magnetic field is governed by the underlying integrability or nonintegrability of quantum dots to which an electron gas is confined. Similar considerations will be devoted to the mesoscopic persistent current I caused by the Aharonov–Bohm effect.

6.1 Historical Background

The theory and experiment of **orbital diamagnetism** associated with the cyclotron motion have almost a century of history. A charged particle in a magnetic field, receiving a Lorentz force, obeys cyclotron motion, which is a basic principle for accelerators. The orbital diamagnetism also provides a typical phenomenon in which to observe a quantum manifestation of chaos. In the presence of a homogeneous magnetic field (e.g., parallel to the z direction) $\boldsymbol{B}(=B\boldsymbol{n}_z)$, an electron with mass m_e and charge $-e$ is described by the equation of motion

$$m_e \frac{d^2\boldsymbol{r}}{dt^2} = -\frac{e}{c}\boldsymbol{v} \times \boldsymbol{B}, \tag{6.1}$$

where c is the velocity of light. The electron shows cyclotron motion (Larmor orbit) around the $\boldsymbol{B}$ field, as well as translational motion parallel to the field. The cyclotron motion is characterized by the cyclotron radius $R_c = \dfrac{m_e v c}{eB}$ and frequency $\omega_c = \dfrac{eB}{m_e c}$. This rotational component of the orbital motion generates a loop current, yielding the diamagnetic moment $\mu = \dfrac{eR_c^2}{2c}\omega_c$ that is antiparallel to the $\boldsymbol{B}$ field. Hence the name of diamagnetism arises.

Now, what is the nature of diamagnetism for the elecron gas confined to a metal? Let us apply classical statistitical mechanics to the interacting electron gas in the $\boldsymbol{B}$

field. With the use of the vector potential $\boldsymbol{A}$ $(\boldsymbol{B} = \nabla \times \boldsymbol{A})$, the Hamitonian for the electron gas is given by[8]

$$\mathcal{H} = \sum_j \frac{1}{2m_e} \left\{ \boldsymbol{p}_j + \frac{e}{c} \boldsymbol{A}(\boldsymbol{r}_j) \right\}^2 + U(\{\boldsymbol{r}_j\}). \tag{6.2}$$

The partition function for this system is

$$Z = \int \prod_j d^3 r_j \int \prod_j d^3 p_j \exp\left(-\frac{1}{kT}\mathcal{H}\right). \tag{6.3}$$

The change of the integration variable from $\boldsymbol{p}$ to $\boldsymbol{\pi}$ $\left(= \boldsymbol{p} + \dfrac{e}{c}\boldsymbol{A}\right)$ will sweep $\boldsymbol{A}$ away in Z. In consequence, the free energy $F = -kT \ln Z$ has no external parameter B, resulting in the vanishing susceptibility $\left(\chi = -\dfrac{d^2 F}{dB^2} = 0\right)$. This result, however, is inconsistent with the non-zero diamagnetic susceptibility observed in experiments!

In the thesis for his doctorate, Bohr pointed out the above mystery in 1911. Ten years later, Ms. van Leeuwen made[67] an interesting observation: Noting the finite size of real metals, she addressed the emergence of an edge current caused by electrons skipping along the metal boundary, surmising that this current would cancel the bulk current coming from the complete cyclotron (Larmor) orbits (see Fig. 6.1). As illustrated in Fig. 6.1, four Larmor orbits in the center generate a counter-clockwise bulk current, whereas the skipping orbit along the metal boundary leads to the clockwise edge current. The net current in the metal thus vanishes. Nevertheless, her idea failed in obtaining a finite diamagnetic susceptibility. Although a complete explanation of the diamagnetism would have to wait for the discovery of quantum mechanics four years later, the idea of edge currents deserves further intensive consideration and Ms. van Leeuwen, who conceived the novel transport via electrons colliding with metal boundaries, should be counted among the pioneers with Kelvin, Krylov, Sinai, and others[31-36], in the fields of **nonlinear dynamics and chaos in billiards**.

In 1930 after the birth of quantum mechanics, Landau solved the (time-independent) Schrödinger equation for an electron in a magnetic field under the condition that the wavefunction should vanish at infinity. With the use of the eigenvalues or the so-called **Landau levels**

$$E_n = \left(n + \frac{1}{2}\right)\hbar\omega_c \qquad (n = 0, 1, \cdots) \tag{6.4}$$

as single-electron levels, Landau developed quantum statistical mechanics for an electron gas, deriving successfully the diamagnetic susceptibility

---

[8]From the Lagrangian $\mathcal{L} = \frac{1}{2}m_e \dot{\boldsymbol{r}}^2 - \frac{e}{c}\dot{\boldsymbol{r}} \cdot \boldsymbol{A}(\boldsymbol{r})$, the canonical momentum $\boldsymbol{p}$ is defined as $\boldsymbol{p} = \frac{\partial \mathcal{L}}{\partial \dot{\boldsymbol{r}}} = m_e \dot{\boldsymbol{r}} - \frac{e}{c}\boldsymbol{A}(\boldsymbol{r})$, and the Hamiltonian is $\mathcal{H} = \boldsymbol{p}\dot{\boldsymbol{r}} - \mathcal{L} = \frac{1}{2m_e}\left\{\boldsymbol{p} + \frac{e}{c}\boldsymbol{A}(\boldsymbol{r})\right\}^2$.

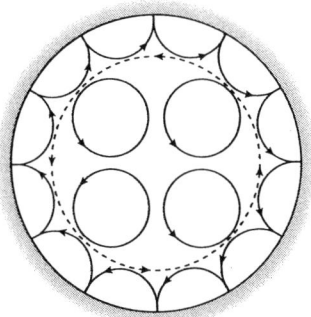

Fig. 6.1   van Leeuwen's bulk current and edge current.

$$\chi^{\text{Landau}} = -\frac{1}{3}\mu_B^2 g(E_F),$$

where $g(E_F)$ is the density of states at the Fermi energy E_F, and $\mu_B = \dfrac{e\hbar}{2m_e c}$ is the Bohr magneton. The value $-\chi$ takes a maximum at $B = 0$, showing de Haas–van Alphen oscillation as B is increased. By the way, Landau's theory ignores the boundary effect or the underlying nonlinear electron dynamics that is inevitably encountered in mesoscopic billiards. Once we take into consideration these effects, Landau levels and their offspring become insufficient, and a qualitatively new insight as cultivated by van Leeuwen is demanded, which is nothing but a theme of quantum chaos.

There is a related subject of the persistent current (PC). Let us consider a mesoscopic ring (doughnut) with the magnetic flux penetrating through its hollow. One can expect PC along the ring caused by the Aharonov–Bohm effect. (Details will be described later.) In fact, Mailly *et al.*[66] observed PC in a ballistic GaAs/Al$_x$Ga$_{1-x}$As ring. This current is sensitive to the boundary geometry and the underlying nonlinear dynamics. Thus, PC as well as diamagnetism constitute subjects of quantum chaos.

6.2   Orbital Diamagnetism in the Light of Nonlinear Dynamics

Reflecting the growing attention to chaos, the study of the nonlinear dynamics of an electron in a magnetic field was kicked off by Robnik and Berry[68]. Below, we shall summarize their findings. Suppose an electron is confined to the 2-d elliptic billiard system with longer radius a and shorter one b, and perimeter Λ. In the presence of the magnetic field B perpendicular to the ellipse, the electron dynamics is represented geometrically by a series of incomplete arcs characterized by the Larmor radius R_c (see Fig. 6.2). Let the position at the boundary where an electron bounces be denoted by the length along the ellipse boundary s (mod Λ) and the cosine of the reflection angle by $p = \cos\theta$: (s, p) constitute the Birkhoff coordinates (see Chapter 3). The electron trajectory is now described by using the 2-d discrete map for Birkhoff coordinates (s_n, p_n) generated through successive collisions $(n = 1, 2, \cdots)$ with the wall:

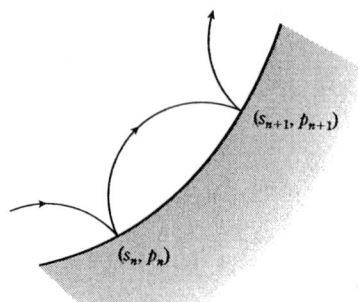

Fig. 6.2   Dynamical variables related to successive collisions.

$$s_{n+1} = F_1(s_n, p_n),$$
$$p_{n+1} = F_2(s_n, p_n). \tag{6.5}$$

We define three characteristic radii, one for the largest inscribed circle R^*, another for the largest curvature $\rho_{\min}$, and the other for the smallest curvature $\rho_{\max}$. They are explicitly written as

$$R^* = b, \qquad \rho_{\min} = \frac{b^2}{a}, \qquad \rho_{\max} = \frac{a^2}{b},$$

satisfying $\rho_{\min} < R^* < \rho_{\max}$. As the Larmor radius R_c is decreased, the phase-space structure shows a radical change: (1) in the weak-field case ($\rho_{\max} < R_c < \infty$), each orbit resembles a sequence of straight linear segments, but the major part of the phase space is occupied by the chaotic sea; (2) in the intermediate-field case ($\rho_{\min} < R_c < \rho_{\max}$), the chaos arises only from diametrical orbits along the long diameter of the ellipse and this chaotic fraction diminishes with decreasing R_c; (3) in the strong-enough field case ($R_c < \rho_{\min}$), van Leeuwen's stable skipping motions along the billiard table boundary dominate the phase space. Precisely speaking, in addition to orbits described by the collision map (6.5), the stable complete Larmor orbits that do not touch the wall will coexist in the case $R_c < R^*$.

Numerical iteration of (6.5) leads to the Poincaré surface of section in Fig. 6.3, which conveys detailed information on the trajectories. In the zero-field limit $R_c = \infty$, the system is integrable with its phase space occupied by the **invariant tori (KAM tori)** (see Fig. 6.3(a)). In Fig. 6.3(a), two families of tori can be found, one encircling around each of the *eyes* and the other drifting along the s axis. The border curve between these two is called the **separatrix**, which is most unstable against a change of the B field. In the weak and intermediate field case ($\rho_{\min} < R_c < \infty$) in Figs. 6.3(b) and 6.3(c), orbits around the separatrices become chaotic and the chaotic fraction of the phase space varies with increasing B field, although one can see no fully developed chaos. In the strong-field case ($R_c < \rho_{\min}$) in Fig. 6.3(d), chaos disappears and the phase space is fully occupied by the invariant tori. By the way, what is the quantal manifestation of the transition from chaos to tori at the critical point $R_c = \rho_{\min}$?

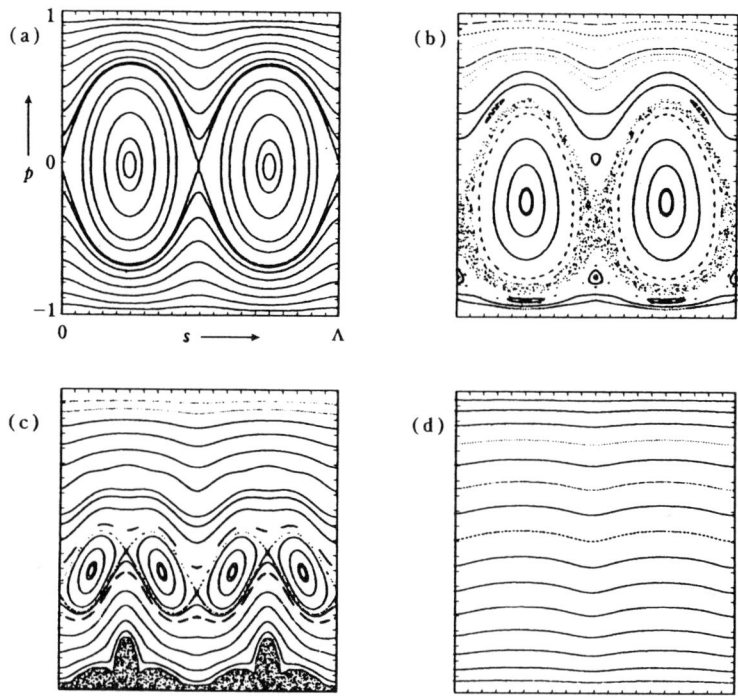

Fig. 6.3   Poincaré surface of section: (a) $R_c = \infty$; (b) $R_c = 3$; (c) $R_c = 0.75$; (d) $R_c = 0.4$. $s = 0$ and $s = \dfrac{\Lambda}{2}$ correspond to both ends of the long diameter (reproduced from[68]).

This question was first answered by Nakamura and Thomas[69]. Let us quantize a 2-d noninteracting elecron gas within the elliptic billiard system in the applied $\boldsymbol{B}$ field, and derive a single-electron spectrum. (We ignore the lattice periodicity and spin degree of freedom.) Using the area for the ellipse $\pi S = \pi a b$, we make dimensionless the electron energy $E \left(= \dfrac{1}{2} m v^2 \right)$ and other quantities:

$$\overline{E} = E \left(\frac{\hbar^2}{2mS} \right)^{-1}, \qquad \overline{B} = B \left(\frac{c\hbar}{eS} \right)^{-1},$$
$$\overline{R_c} = R_c S^{-1/2}, \qquad \overline{\rho} = \rho S^{-1/2}.$$

According to the quantal–classical correspondence, the energy spectra in the region satisfying

$$\left(\frac{\overline{E}}{\overline{B}^2}\right)^{1/2} (= \overline{R}_c) > \overline{\rho}_{\min}$$

should show a feature different from the one in the opposite region.

Figure 6.4(a) shows single-particle energies as a function of the B field. In the strong-field region $\overline{B} \gg 1$, bunchings to Landau levels are evident. However, in the lower-B field, avoided crossings (level repulsions) among energy levels are prominent. The broken circular symmetry in the ellipse has changed true crossings among differnt angular-momentum manifolds to avoided crossings. Careful insight reveals a multitude of avoided crossings with distinguished energy gaps and the resultant complicated energy spectra in the lower-B and higher-E region satisfying $R_c > \rho_{\min}$ or equivalently $\frac{\overline{E}}{\overline{B}^2} > \overline{\rho}_{\min}^2$. These high concentrations of avoided crossings with distinguished energy gaps will be a quantum signature of chaos.

To obtain the ground state energy $E_G(B)$ of the electron gas, we must fill the single-particle levels in Fig. 6.4(a) each with two electrons from the bottom up to the Fermi energy $E_F (= E_N)$, following the Fermi statistics. The zero-temperature susceptibility per electron is defined by

$$\chi = -\frac{1}{2N}\frac{d^2 E_G(B)}{dB^2}.$$

Nakamura and Thomas calculated χ numerically by using the difference as

$$\chi = -\frac{1}{N}\frac{\Delta^2}{\Delta B^2}\left[\sum_{1 \le j \le N} E_j\right] = -\frac{2m_e S}{\hbar^2}\frac{\mu_B^2}{N}\sum_{1 \le j \le N}\frac{\Delta^2 \overline{E}_j}{\Delta \overline{B}^2}. \tag{6.6}$$

Since $\dfrac{\Delta^2 \overline{E}_j}{\Delta \overline{B}^2}$ implies the curvature of individual levels, (6.6) is nothing but the negative of the mean curvature. For a given $\overline{E}_F$, wherever $\dfrac{\overline{E}_F}{\overline{B}^2} > \overline{\rho}_{\min}^2$ holds, a multitude of avoided crossings cause cancellation of positive and negative curvatures leading to a greatly reduced magnitude of χ.

Figure 6.4(b) is the numerically calculated result for χ in units of $\dfrac{2m_e S}{\hbar^2}\mu_B^2$. In this unit, noting $g_{2d}(E_F) = \dfrac{m_e S}{\hbar^2}$ the Landau susceptibility (per electron) at zero temperature is

$$-\chi^{\text{Landau}} = \frac{1}{6} \approx 0.166.$$

From this figure, one finds the following:

(i) For an ellipse, $-\chi$ at $B = 0$ has a value much less than the Landau susceptibility and grows as B is increased.

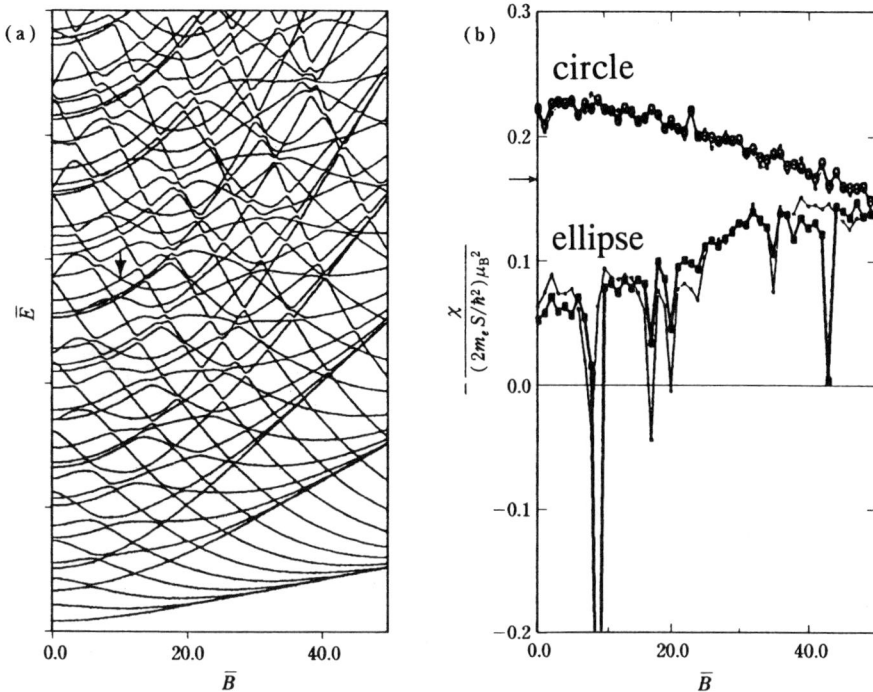

Fig. 6.4 Quantum-mechanical features of an elliptic billiard system $\left(\dfrac{b}{a} = 0.5\right)$: (a) single-electron energies versus magnetic field (even-parity manifold); (b) χ versus magnetic field. The result for the circle $\left(\dfrac{b}{a} = 1\right)$ is also included. The arrow indicates the zero-field Landau susceptibility. The bold lines denote the contribution from both even- and odd-parity manifolds. The thin line is a contribution from the even-parity manifold alone. (Reproduced from[69]).

(ii) For a circle, on the contrary, $-\chi$ at $B = 0$ has a value much larger than the Landau susceptibility and decreases with increasing B. (For a circle, the underlying classical dynamics is integrable showing regular orbits only. The corresponding quantum energy spectrum is monotonic, having no avoided crossing. As a result, all energy levels have positive curvatures giving their large mean value.)

(iii) There is a critical field B_c that satisfies $\dfrac{\overline{E}_F}{\overline{B}_c^2} = \overline{p}_{\min}^2$. For $B > B_c$, the features of χ coincide for both the ellipse and the circle. The conclusion on the ellipse (circle) that the magnitude of the susceptibility is less (larger) than the Landau susceptibility is not available from Landau's theory itself and is a quantal manifestation of chaos (integrability)[22].

Although the above discovery is interesting, the results were derived on the basis of quantum numerical calculations. Fluctuations of curvatures associated with sharp avoided crossings cannot correctly be evaluated by means of the difference in (6.6) and this has an inevitable limitation on resolution. The size of the energy matrix is not large enough, which makes it difficult to surmise the asymptotic behavior of χ as the electron number or Fermi energy is inceased. The finite temperature effect on χ also remains unclear. All these clouds will be swept away by having recourse to the semiclassical theory. Choosing 3-d billiards in the weak magnetic field, we shall now proceed to the semiclassical analysis of the diamagnetism (see Ma and Nakamura [76,77]). For 2-d billiards, see Oppen[72] and Richter, Ullmo and Jalabert[60].

6.3 Semiclassical Orbital Diamagnetism in 3-d Billiards

Three-dimensional (3-d) billiards – systems in which a particle moves inside a closed wall surface – are the simplest among dynamical systems with three degrees of freedom. Three-dimensional billiards have the advantage that they are free from the Anderson localization phenomena proper to low-dimensional ($d = 1$ and 2) systems in a diffusive regime. The theory below is valid in the ballistic regime where electrons obey a deterministic law rather than the diffusion equation. If a spherical shell of sub-micron scale were fabricated by exploiting drying etching processes with focused-ion beams applied to $Al_x Ga_{1-x} As/GaAs/Al_x Ga_{1-x} As$ double heterostructures, it will be smaller than both the elastic mean free path ($\sim 10\,\mu m$) and phase coherence length ($> 10\,\mu m$), ensuring the ballistic regime.

We consider two types of 3-d billiards, (a) a **completely integrable** system and (b) a **fully chaotic** system. The definition of complete integrability (see Section 5.1) tells us that the system (a) has nontrivial constants of motion besides the *energy*. Consider, for instance, a **spherical shell** with $SO(3)$ symmetry (see Fig. 6.5).

The nontrivial constants of motion in this system, namely, two components of the angular momentum, are characterized by means of continous parameters. According to **Liouville's theorem**[1], the system with the number of independent constants of motion equal to the degrees of freedom is completely integrable. On the contrary, the system (b) has no constant of motion other than the energy. As a concrete example, a 3-d billiard system with no continuous symmetry will be taken. In what follows, we shall apply the semiclassical theory to these 3-d mesoscopic billiards, showing how the susceptibility should differ between the types (a) and (b).

In isolated mesoscopic billiards, the electron number N is fixed. The susceptibility at finite temperature in such a system can be obtained by using the framework of the canonical ensemble as (see Chapter 5)

$$\chi = -\frac{1}{V} \left(\frac{\partial^2 F}{\partial B^2} \right)_{T,N},$$

$$(6.7)$$

where $F (= \mu N + \Omega(T, B, \mu))$, V and T are the free energy, volume and temperature, respectively. The thermodynamic potential Ω in the grand canonical ensemble

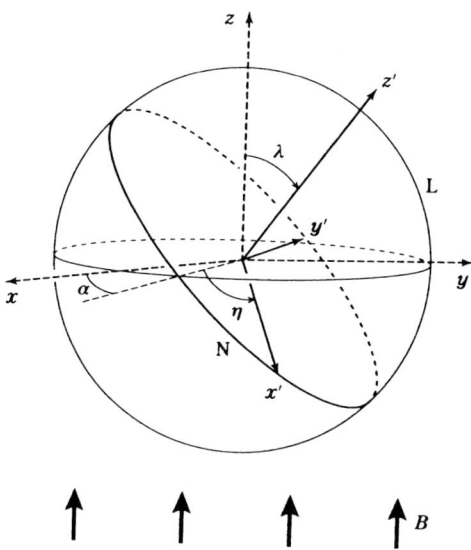

Fig. 6.5   Spherical shell billiard system and its great circle N. For simplicity, the inner sphere is suppressed. α, η, λ are Euler angles.

is defined in terms of the chemical potential μ, the inverse temperature $\beta = \dfrac{1}{k_B T}$ and the density of states $g(E)$, by

$$\Omega(T, B, \mu) = -\frac{1}{\beta} \int_0^\infty dE\, g(E) \ln\left(1 + \exp[\beta(\mu - E)]\right). \tag{6.8}$$

As emphasized in Chapter 5, $g(E)$ is a sum of the mean $\bar{g}(E)$ and oscillating $\delta g(E)$ parts. At low temperatures higher than the mean level spacing but less than μ, the free energy can be expanded (see Section 5.2). With help of $\bar{g}(E)$, let us define the mean chemical potential $\bar{\mu}$ that fixes the value N. Using an expansion for the chemical potential, $\mu \cong \bar{\mu} + \delta\mu$, F is expanded up to a term bilinear in $\delta\mu$ as

$$F \cong F^{(0)} + \delta F^{(1)} + \delta F^{(2)},$$

where each term is given by

$$F^{(0)} = \bar{\mu} N + \overline{\Omega}(\bar{\mu}), \tag{6.9a}$$

$$\delta F^{(1)} = \sum_{po} \left(\frac{\hbar v_F}{L_{po}}\right)^2 \delta g_{po}(\bar{\mu}) \equiv \left(\frac{L_{po}}{L_T}\right), \tag{6.9b}$$

$$\delta F^{(2)} = \frac{(\delta N)^2}{2\bar{g}(\bar{\mu})} = \frac{1}{2\bar{g}(\bar{\mu})} \left[\sum_{po} \frac{\hbar v_F}{L_{po}} \delta g_{po}^\sharp(\bar{\mu}) \equiv \left(\frac{L_{po}}{L_T}\right)\right]^2. \tag{6.9c}$$

In the above, $\Xi\left(\dfrac{L_{po}}{L_T}\right) = \dfrac{\dfrac{L_{po}}{L_T}}{\sinh\left(\dfrac{L_{po}}{L_T}\right)}$ is a thermal damping factor with the cutoff

length $L_T = \dfrac{\hbar v_F \beta}{\pi}$, δN is the oscillating part of the particle number, and the symbol

$\sharp$ prescribes a phase shift by $-\dfrac{\pi}{2}$ (see Chapter 5). $\delta F^{(2)}$ in (6.9c) is an important term proper to mesoscopic systems. A **typical susceptibility** χ for individual billiards is derived by the second-order differential of $-(F^{(0)} + \delta F^{(1)})$ with respect to a magnetic field B. In real experiments, a lot of single quantum dots are fabricated on a given plane, so that one is concerned with the **ensemble-averaged value** $\overline{\chi}$ that is derived from $\delta F^{(2)}$. Below we concentrate on low temperatures of practical interest, $\mu \gg k_B T \gg \hbar\omega_c$. ($\omega_c$ is the cyclotron frequency.)

First of all we confirm the expression for the Landau susceptibility within a semi-classical framework:

$$\chi^{\mathrm{Landau}} = -\frac{\mu_B^2}{3V}\overline{g} \tag{6.10}$$

with μ_B the Bohr magneton. In (6.10), $\overline{g}$ is the universal mean density of states for 3-d billiards, and is determined by only the volume V and surface area S of the billiard system (see Chapter 5) as[13]

$$\overline{g} = \frac{1}{4\pi^2}\left(\frac{2m_e}{\hbar^2}\right)^{3/2}\sqrt{E}\,V - \frac{1}{16\pi}S. \tag{6.11}$$

For a 3-d billiard system with Fermi energy $E = E_F = \dfrac{\hbar^2 k_F^2}{2m_e}$, a combination of (6.10) and (6.11) together with the definition for $\mu_B\ (= \dfrac{e\hbar}{2m_e c}$) yields

$$\chi_{3d}^{\mathrm{Landau}} = -\frac{g_s e^2 k_F}{24\pi^2 m_e c^2} \tag{6.12}$$

with $g_s\ (= 2)$ a spin factor. In d-dimensional billiards in general, we have $\chi_d^{\mathrm{Landau}} \propto k_F^{d-2}$.

As explained in Chapter 5, $\overline{g}$ is a universal quantity common to both integrable and chaotic systems. On the other hand, the oscillating part δg depends on the individual nature of the billiards. For completely integrable systems, the Berry–Tabor trace formula should hold, unless continuous symmetry is present and periodic orbits are degenerate. However, a large number of integrable systems possess continuous symmetry, for which improved versions of the trace formula are proposed. The completely integrable spherical shell billiards (a) have non-Abelian $SO(3)$ continuous symmetry. To such a system, we shall apply the Creagh–Littlejohn trace formula (5.35) in Chapter 5 which extends Gutzwiller's to an integrable system with continuous symmetry. On the contrary, in the fully chaotic billiards (b) where all periodic orbits are isolated, the original Gutzwiller's trace formula is a powerful tool.

6.3.1 Integrable (Spherical Shell) Billiards

Here we shall first derive the zero-field density of states, and then move on to a perturbational treatment to obtain the density of states for a system in the weak B field. Let **spherical shell billiards** have outer (radius R) and inner (radius r) spheres where an electron gas is confined to the shell. The system with $SO(3)$ symmetry has $f = 3$ independent continuous parameters represented by the Euler angles. The trace formula applicable to such a system is obtained from (5.35) (Creag and Littlejohn [54, 55])

$$\delta g(E) = \sum_\Gamma \delta g_\Gamma(E)$$

$$= \frac{1}{\pi\hbar} \frac{1}{(2\pi\hbar)^{3/2}} \sum_\Gamma \frac{8\pi^2 J_\Gamma T_\Gamma}{n_\Gamma \left|\left(\dfrac{\partial\theta_{//}}{\partial J_\Gamma}\right)\right|^{1/2}} \cos\left(\frac{1}{\hbar}S_\Gamma - \mu_\Gamma\frac{\pi}{2} - \frac{3}{4}\pi\right).$$

$$(6.13)$$

This is the zero-field result. In (6.13) Γ represents each family of degenerate periodic orbits (PO's) with n_Γ the internal discrete-rotational symmetry of PO; S_Γ, μ_Γ and T_Γ are the action, Maslov index and period, respectively. The phase shift $\left(-\frac{3}{4}\pi = -3 \times \frac{\pi}{4}\right)$ comes from $f = 3$ independent parameters, and J_Γ and $\theta_{//}$ stand for the magnitude of the angular momentum and total rotation angle, respectively. In the pre-cosine factor in (6.13), $\dfrac{8\pi^2 J_\Gamma T_\Gamma}{n_\Gamma}$ and $\dfrac{\partial\theta_{//}}{\partial J_\Gamma}$ come from integration over the group measure and the stability of one typical orbit of each Γ, respectiverly, in (5.35).

As an electron obeys specular reflection at the spherical surface, its periodic orbits (PO's) lie on the **great circular plane (GCP)** sharing the same origin as that of the double spheres (see Fig. 6.5). The GCP is specified by Euler angles (α, η, λ) with λ the angle between the normal vector of GCP and the z axis. The present system has two kinds of PO's, one bouncing at the surface of both the outer and inner spheres (type I) and the other bouncing at the surface of the outer sphere only (type II) (see Fig. 6.6). The following procedure is to calculate S_Γ and other quantities for each periodic orbit, Γ, and to substitute them into (6.13).

Now we give concrete expressions in the case of type I orbits, Γ_1. Using the number of bouncings with the outer surface (v_1), length of PO (L_{Γ_1}), and Fermi momentum (velocity) ($p_F(v_F)$), one finds

$$J_{\Gamma_1} = p_F R \sin\alpha_1, \qquad (6.14)$$

$$T_{\Gamma_1} = \frac{L_{\Gamma_1}}{v_F} = \frac{2v_1}{v_F}\sqrt{R^2 + r^2 - 2Rr\cos\left(\frac{\pi}{v_1}\right)}, \qquad (6.15)$$

$$S_{\Gamma_1} = 2p_F v_1 \sqrt{R^2 + r^2 - 2Rr\cos\left(\frac{\pi}{v_1}\right)}. \qquad (6.16)$$

α_1 in (6.14) denotes the angle of reflection at the outer surface and is given by

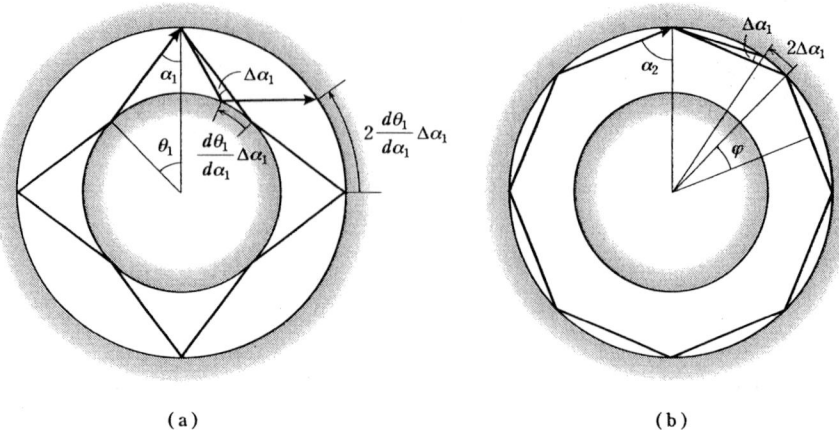

Fig. 6.6   Periodic orbits in the great circular plane (GCP) of spherical shell billiards: (a) type I orbit; (b) type II orbits.

$$\alpha_1 = \arcsin \left[\frac{r \sin \left(\dfrac{\pi}{v_1} \right)}{\sqrt{R^2 + r^2 - 2Rr \cos \left(\dfrac{\pi}{v_1} \right)}} \right]. \tag{6.17}$$

Further, let θ_1 be the rotation angle for each bounce with the external surface. Then for PO (Γ_1) the sensitivity of the total rotation angle $\theta_{/\!/} = 2v_1\theta_1$ to the angular momentum J is quantified as

$$\left(\frac{\partial \theta_{/\!/}}{\partial J} \right)_{\Gamma_1} = \frac{2v_1}{p_F R \cos \alpha_1} \left(\frac{d\theta_1}{d\alpha_1} \right)_{\Gamma_1}, \tag{6.18}$$

where

$$\theta_1 = \arcsin \left[\frac{1}{r} \left(\frac{R \sin(2\alpha_1)}{2} - \sqrt{r^2 \sin^2 \alpha_1 - R^2 \sin^4 \alpha_1} \right) \right]. \tag{6.19}$$

Equations (6.14)–(6.19) provide the quantities sufficient to compute (6.13). The corresponding ones in the case of type II orbits, Γ_2, are available in a similar way.

The effect of a weak B field can be incorporated perturbatively, namely, by adding to S_Γ in (6.13) a correction

$$\Delta S_{\Gamma(\alpha,\eta,\lambda)} = \frac{e}{2\pi c} B\Theta_\Gamma \cos \lambda, \tag{6.20}$$

where Θ_Γ is $2\pi\times$ area enclosed by PO (Γ) and is premised to be positive for clockwise rotations. Given GCP's specified by Euler angles in the range $(\alpha, \eta, \lambda) \sim (\alpha + d\alpha, \eta +$

$d\eta, \lambda + d\lambda$), an assemby of PO's lying on those planes give a contribution to the density of states as

$$d\delta g|_{\alpha, \eta, \lambda} = \frac{1}{8\pi^2} \sin \lambda \, d\alpha d\eta d\lambda \sum_\Gamma \cos \left(\frac{1}{\Phi_0} B \Theta_\Gamma \cos \lambda \right) \delta g_\Gamma (B = 0),$$

(6.21)

where Γ includes both type I and II orbits, $\Phi_0 = \dfrac{hc}{e}$ is the flux quantum, and $\delta g_\Gamma (B = 0)$ means the zero-field value. Also, noting the time-reversal symmetry in the zero-field limit, the relations $S_{-\Gamma} = S_\Gamma, \Theta_{-\Gamma} = -\Theta_\Gamma$, etc. have been exploited.

Integration of (6.21) over the Euler angles brings out the oscillating part of the density of states,

$$\delta g(B) = \sum_\Gamma j_0 \left(\frac{1}{\Phi_0} B \Theta_\Gamma \right) \delta g_\Gamma (B = 0) = \sum_\Gamma j_0 \left(\overline{\Phi} \frac{\Theta_\Gamma}{\pi R^2} \right) \delta g_\Gamma (B = 0).$$

(6.22)

Here $j_0 \left(j_0(x) = \dfrac{\sin x}{x} \right)$ is the spherical Bessel function of order zero, and $\overline{\Phi} = \dfrac{\pi B R^2}{\Phi_0}$ is the dimensionless magnetic flux scaled by Φ_0 penetrating the external sphere. The amplitude factor in $\delta g_\Gamma (B = 0)$ is approximately determined by noting

$$T_\Gamma \sim \frac{L_\Gamma}{v_F}, \qquad J_\Gamma \sim p_F R, \qquad \left(\frac{\partial \theta_{//}}{\partial J} \right)_\Gamma \sim \frac{1}{p_F R}.$$

Substituting (6.22) into the formula $\chi = -\dfrac{1}{V} \left(\dfrac{d^2 \delta F^{(1)}}{dB^2} \right)$, one reaches the typical susceptibility for individual spherical shell billiards written in dimensionless form ($F^{(0)}$, having no B-dependent correction, does not contribute to χ):

$$\frac{\chi}{\chi_0^A} = -g_s \left(\frac{R^3}{V} \right) \sqrt{\frac{2}{\pi}} (k_F R)^{5/2} \sum_\Gamma \frac{R}{L_\Gamma} \Xi \left(\frac{L_\Gamma}{L_T} \right)$$

$$\times \frac{d^2 j_0 \left(\overline{\Phi} \frac{\Theta_\Gamma}{\pi R^2} \right)}{d \overline{\Phi}^2} \cdot \frac{\sin \alpha_\Gamma \cos \left(\frac{1}{\hbar} p_F L_\Gamma - \mu_\Gamma \frac{\pi}{2} - \frac{3}{4} \pi \right)}{n_\Gamma \left| p_F R \left(\frac{\partial \theta_{//}}{\partial J} \right)_\Gamma \right|^{1/2}},$$

(6.23)

where α_Γ as exemplified in α_1 and α_2 (see Fig. 6.6) denotes the angle of reflection at the outer surface, and $\chi_0^A = \dfrac{e^2}{m_e c^2 R}$ is the unit of susceptibility. Owing to the thermal damping factor Ξ, only short PO's with $L_\Gamma \sim 2\pi R, \Theta_\Gamma \sim \pi R^2$ can survive.

From (6.23), we find that the **typical susceptibility can be both paramagnetic and diamagnetic:**[9] As a function of k_F at fixed B, χ exhibits a growing

---

[9]The induced magnetic moment is called paramagnetic (diamagnetic) when it is parallel (antiparallel) to the applied field.

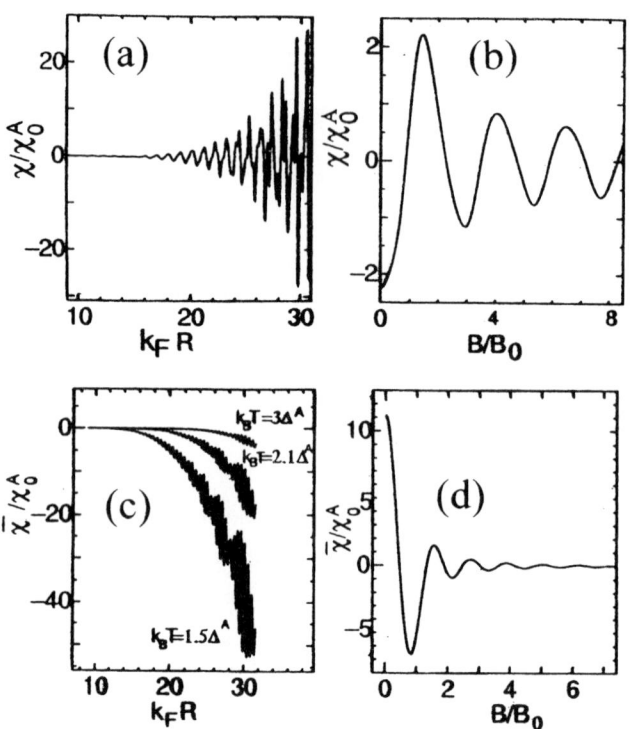

Fig. 6.7   Susceptibilities for a spherical shell, scaled by χ_0^A. $r/R = 0.90$, $T = 1.50\Delta^A/k_B$: (a) χ versus $k_F R$ at $B = B_0$; (b) χ versus B at $k_F R = 19.0$; (c) $\overline{\chi}$ versus $k_F R$ at $B = B_0$; (d) $\overline{\chi}$ versus B at $k_F R = 20.0$. Figure 6.7(c) includes results at $T = 2.1\Delta^A/k_B$ and $T = 3.0\Delta^A/k_B$ (reproduced from[76]).

oscillation with period $\sim R^{-1}$ and its magnitude is $\chi \propto k_F^{5/2}$, which is much larger than $\chi_{3d}^{\text{Landau}}$ ($\propto k_F^1$). The function j_0 determines the B-dependent feature. χ, as a function of B at fixed k_F, shows a damping oscillation with period $\sim \dfrac{\Phi_0}{\pi R^2}$ and its magnitude is inversely proportional to B. The results are displayed in Figs. 6.7(a) and 6.7(b) for a single spherical shell with $r/R = 0.90$ at temperature $T = 1.5\Delta^A/k_B$.

The ensemble-averaged susceptibility $\overline{\chi}$ – another important observable – is obtained from $\delta F^{(2)}$ in (6.9c). As described in Section 5.2, $(\delta N)^2$ averaged over the Fermi energy or sample size gives a nonvanishing value, $\langle(\delta N)^2\rangle$, which comes from the diagonal terms only. Using this result in the formula

$$\overline{\chi} = -\frac{1}{2V\overline{g}}\frac{\partial^2}{\partial B^2}\langle \delta N^2(\mu, B)\rangle,$$

one finds

$$\frac{\overline{\chi}}{\chi_0^A} = -g_s \left(\frac{R^3}{V}\right)^2 2\pi (k_F R)^2 \sum_\Gamma \Xi^2 \left(\frac{L_\Gamma}{L_T}\right)$$

$$\times \frac{d^2 j_0^2 \left(\overline{\Phi}\frac{\Theta_\Gamma}{\pi R^2}\right)}{d\overline{\Phi}^2} \cdot \frac{\sin^2 \alpha_\Gamma \cos^2 \left(\frac{1}{\hbar}p_F L_\Gamma - \mu_\Gamma \frac{\pi}{2} - \frac{3\pi}{4}\right)}{n_\Gamma^2 \left|p_F R \left(\frac{\partial \theta_{//}}{\partial J}\right)_\Gamma\right|}. \tag{6.24}$$

(6.24) indicates that whether the ensemble-averaged susceptibility is paramagnetic or diamagnetic is determined by the B field: at fixed k_F, $\overline{\chi}$ takes its maximum paramagnetic value at $B = 0$ and oscillates around $\overline{\chi} = 0$ with respect to B with the magnitude of its oscillation declining as B^{-2}; at fixed B, $\overline{\chi} \propto k_F^2$, which is as yet larger than $\chi_{3d}^{\text{Landau}}$. Figures 6.7(c) and 6.7(d) confirm these results.

6.3.2 Fully Chaotic Billiards

Now, let us move forward to the susceptibility for chaotic systems. In fully chaotic 3-d billiards, all the PO's are isolated from each other, ensuring the applicability of the Gutzwiller trace formula (see Chapter 5):

$$\delta g(E) = \sum_\Gamma \delta g_\Gamma(E)$$

$$= \frac{1}{\pi\hbar} \sum_\Gamma \frac{T_\Gamma}{\left|\det(\tilde{M}_\Gamma - I)\right|^{1/2}} \cos\left(\frac{1}{\hbar}S_\Gamma - \mu_\Gamma \frac{\pi}{2}\right). \tag{6.25}$$

In the presence of a weak B field, the perturbation theory holds as in the previous case (a). By adding the field-dependent correction to the action, (6.25) becomes

$$\delta g(B) = \sum_\Gamma \cos\left(\frac{1}{\Phi_0}\sum_i B\Theta_{\Gamma_i}\right)\delta g_\Gamma(B = 0). \tag{6.26}$$

Θ_{Γ_i} means the area of the projection of the triangle onto the plane perpendicular to the B field with the triangle itself formed by a succession of three collisions with the wall. Let L be the characteristic length of the chaotic billiard system, and then $\sum_i \Theta_{\gamma_i} \sim L^2$ for the shortest PO's (γ) that survive at low but finite temperatures.

The dimensionless flux, now defined as $\overline{\Phi} = \dfrac{BL^2}{\Phi_0}$, reduces (6.26) to

$$\delta g(B) \sim \sum_\gamma \cos\left(\overline{\Phi}\sum_i \frac{\Theta_{\gamma_i}}{L^2}\right)\delta g_\gamma(B = 0).$$

The integration of the cosine function is not made here because of the isolated nature of the PO's, which is distinct from the previous case (a). Noting $T_\gamma \sim \dfrac{L_\gamma}{v_F}$, the typical susceptibility is now

$$\frac{\chi}{\chi_0^B} \propto -(k_F L) \frac{d^2}{d\overline{\Phi}^2} \sum_\gamma \cos\left(\overline{\Phi} \sum_i \frac{\Theta_{\gamma_i}}{L^2}\right) \tag{6.27}$$

in units of $\chi_0^B = \dfrac{e^2}{m_e c^2 L}$. The ensemble-averaged value is

$$\frac{\overline{\chi}}{\chi_0^B} \propto -(k_F L)^{-1} \frac{d^2}{d\overline{\Phi}^2} \sum_\gamma \cos^2\left(\overline{\Phi} \sum_i \frac{\Theta_{\gamma_i}}{L^2}\right). \tag{6.28}$$

Thus the semiclassical features of chaotic billiards are as follows: $\chi \propto k_F$ and $\overline{\chi} \propto k_F^{-1}$, namely, χ is of the same order as χ_{3d}^{Landau} whereas $\overline{\chi}$ is much less than the latter. As for the B dependence, both χ and $\overline{\chi}$ are periodic **without decay**, consisting of a finite number of trigonometric functions.

The experiment by Lévy et al.[65] on an array of quantum dots fabricated on a GaAs heterojunction yielded a huge paramagnetic peak of the ensemble-averaged magnetic susceptibility at $B = 0$, which decreases on a scale of one flux quantum through each dot. The result is consistent with the present theoretical prediction for spherical shell billiards. However, Lévy et al.[65]'s dots are of square geometry in two dimensions. It is desirable to observe experimentally the orbital diamagnetism of both integrable and chaotic billiards in three as well as two dimensions.

6.4 Semiclassical Persistent Current in 3-d Shell Billiards

The nature of the persistent current (PC) lies in the Aharonov–Bohm effect. Suppose a ring is enclosing a thin magnetic flux. Once the angle variable at the circumference is taken as a dynamical variable, electron motion along the ring is equivalent to an infinite linear dynamical system with a periodic potential. Due to this equivalence, the magnetic flux plays the role of additional momentum, causing PC. This argument indicates that a multiply connected structure of the ring is essential. The 3-d spherical shell billiards in the weak magnetic field, however, have no such structure. Nevertheless, PO's in the shell certainly enclose an effective flux and lead to PC.

Figure 6.8 again illustrates completely integrable spherical shell billiards in the presence of a weak field. PO's lie in the GCP. For a given GCP (denoted as N) with Euler angles (α, η, λ), a typical PC along the ring on N is given by

$$I = -c\frac{\partial \delta F^{(1)}}{\partial \Phi} \tag{6.29}$$

with $\delta F^{(1)}$ as given in (6.9b). The flux $\Phi (= \pi R^2 B \cos \lambda)$ and the PO's in (6.9b) are defined for GCP (N) (see Fig. 6.8). To be explicit, $\delta F^{(1)}$ is

$$\delta F^{(1)} = \sum_\Gamma \Xi \left(\frac{L_\Gamma}{L_T}\right) \left(\frac{\hbar v_F}{L_\Gamma}\right)^2 d\delta g \Bigg|_{\alpha,\eta,\lambda} \tag{6.30}$$

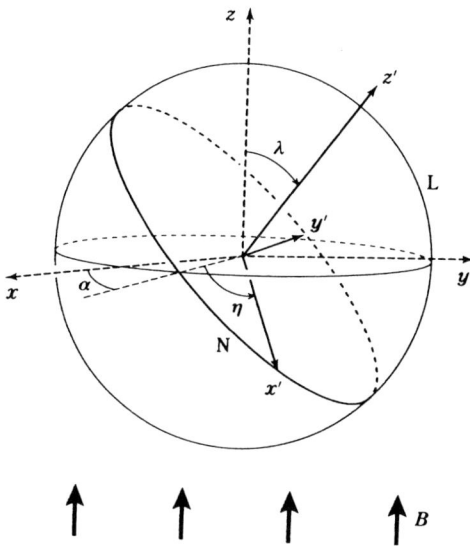

Fig. 6.8   Spherical shell billiards and great circular planes N and M. The inner sphere is suppressed for simplicity. Three tangential vectors at the point H are directed, from above, to one GCP (M), the ring L and another GCP (N).

with $d\delta g|_{\alpha,\eta,\lambda}$ being the same as in (6.21). The PC flowing on N is thus determined by (6.29) and (6.30). Differentiation with respect to the flux through N before integration over Euler angles is a procedure distinct from the one for calculation of the susceptibility.

To obtain the whole PC, we horizontally slice up the spherical shell into thin disks, calculate the partial current flowing on each disk, and finally sum up the contributions from all the thin disks. To realize this strategy, let us first consider the thin ring L at latitude β. While one may calculate the current anywhere in L due to the circular symmetry around z axis, we choose a point H where the outer circle of L crosses the x–z plane. Suppose a pair of GCP's, N and M, commonly pass the point H and are mirror-symmetric about the x–z plane (see Fig. 6.8). The PC along the ring L is a sum of the projections of the equal currents flowing on N and M, and is given by

$$2\cos\rho \times \text{(current flowing on either N or M)}$$

where $\cos\rho = \dfrac{\cos\lambda}{\cos\beta}$. The GCP with polar angle λ satisfying $\beta < \lambda < \dfrac{\pi}{2}$ passes the point H, so the integration over other Euler angles gives rise to the net PC along the ring L:

$$I(B;\beta) = e\hbar v_F^2 g_s \int_\beta^{\pi/2} d\lambda \frac{\cos\lambda}{\cos\beta} \sin\lambda \left[\sum_\Gamma \frac{\Theta_\Gamma}{\pi R^2} \frac{\sin\left(\frac{1}{\Phi_0}B\Theta_\Gamma\cos\lambda\right)}{L_\Gamma L_T \sinh\left(\frac{L_\Gamma}{L_T}\right)} \delta g_\Gamma \right].$$

(6.31)

(6.31) represents the partial PC along the thin ring with latitude β, the integration of which over β results in the total PC flowing around the spherical shell. Noting

$$\delta g_\Gamma \propto \frac{1}{\Delta^A}\left(\frac{p_F R}{\hbar}\right)^{1/2},$$

we eventually have a typical PC

$$\frac{I}{I_0^A} \propto \frac{m_e v_F^2}{\Delta^A}(k_F R)^{1/2} = 2(k_F R)^{5/2}.$$

(6.32)

In the above, $\Delta^A = \dfrac{\hbar^2}{2m_e R^2}$, and $I_0^A = \dfrac{2e\hbar}{m_e R^2}$ are scaling factors for the energy and current, respectively. The result in (6.32) is larger than that of the general argument ($I \propto k_F^2$) by Oppen and Riedel[71].

A similar perturbative calculation for fully chaotic shell billiards (characteristic length L) in the presence of a weak magnetic field leads to

$$\frac{I}{I_0^B} \propto \frac{m_e v_F^2}{\Delta^B}(k_F L)^{-1} = 2k_F L$$

(6.33)

via the zero-field density of states $\delta g_\Gamma \propto \dfrac{1}{\Delta^B}\left(\dfrac{p_F L}{\hbar}\right)^{-1}$ (see (6.26)). Here $\Delta^B = \dfrac{\hbar^2}{2m_e L^2}$ and $I_0^B = \dfrac{2e\hbar}{m_e L}$ are scaling factors, respectively, for the energy and current of chaotic shell billiards.

The enhanced PC was observed by Mailly et al.[66] for a single 2-d Aharonov–Bohm billiard system fabricated on a GaAs/AlGaAs heterojunction. The typical current as a function of flux has the period of flux quantum (hc/e). This double-circle billiard system with radii of order 1μm is less than both the elastic scattering length ($l_e \sim 11\,\mu$m) and the phase-coherence length ($l_\phi \sim 25\,\mu$m) and larger than the Fermi wavelength ($0.042\,\mu$m). Therefore the system falls in the ballistic and semiclassical regime. To compare the theory, more intensive experiments are desirable on the k_F dependence of PC in chaotic as well as integrable shell billiards in two and three dimensions.

We should also compare the PC in 3-d spherical shell billiards with that in the 2-d ring[74], noting the difference of the level density between them. In the case of a 2-d ring, it is shown that $\delta g \propto \dfrac{1}{\Delta_{2d}}\left(\dfrac{p_F R_{2d}}{\hbar}\right)^{-1/2}$, from which one knows

$$I \propto \frac{m_e v_{\mathrm{F}}^2}{\Delta_{2d}} (k_{\mathrm{F}} R_{2d})^{-1/2}$$

with R_{2d} and $\Delta_{2d} = \dfrac{\hbar^2}{2m_e R_{2d}^2}$, being the radius of the outer circle of rings and the mean level spacing, respectively. So, PC in a 3-d spherical shell is much larger than that $(I \propto k_{\mathrm{F}}^{3/2})$ in the 2-d or quasi-1-d rings that have so far attracted experimentalists, and can be observerd more easily in future experiments.

The orbital diamagnetism and persistent current (PC) are typical quantum phenomena in mesoscopic systems. In this chapter, we have analytically investigated these phenomena within the framework of the semiclassical theory combined with a knowledge of chaos and nonlinear dynamics. The types of trace formula are found to be different depending on the integrability or nonintegrability of billiards. We considered two types of 3-d billiards: (a) spherical shell billiards and (b) fully chaotic billiards. Both typical χ and ensemble-averaged $\overline{\chi}$ susceptibilities can be both paramagnetic and diamagnetic, exhibiting growing oscillations with respect to k_{F}. The magnitude of the oscillation is $\chi \propto k_{\mathrm{F}}^{5/2}$ larger than $\chi_{3d}^{\mathrm{Landau}}$ ($\propto k_{\mathrm{F}}^1$) in case (a), and $\chi \propto \chi_{3d}^{\mathrm{Landau}}$ in case of (b). As for the ensemble-averaged value, $\overline{\chi} \propto k_{\mathrm{F}}^2$ still larger than $\chi_{3d}^{\mathrm{Landau}}$ in the case (a), but $\overline{\chi} \propto k_{\mathrm{F}}^{-1}$ less than $\chi_{3d}^{\mathrm{Landau}}$ in the case (b). Concerning the magnetic field dependence, both χ and $\overline{\chi}$ are periodic in B with and without damping in cases (a) and (b), respectively. The result for the typical PC in the corresponding shell billiards is $I \propto k_{\mathrm{F}}^{5/2}$ and $I \propto k_{\mathrm{F}}^1$ in cases (a) and (b), respectively. Both the diamagnetism and persistent current are enhanced in completely integrable sytems, but weakened in fully chaotic systems. It is amusing that these important analytical results are available by having recourse to semiclassical theory.

7

QUANTUM INTERFERENCE IN SINGLE OPEN BILLIARDS

We shall again treat open quantum dots. As stated heretofore, the playground on which to verify quantum manifestations of chaos is provided by a family of ballistic quantum billiards (quantum dots) with their linear dimension between nano (10^{-9}m) through submicron (10^{-7}m) scales less than the electronic mean free path. In ballistic systems, electrons do not obey diffusive motion (e.g., Brownian motion) governed by scattering with impurities, but rather execute ballistic motion like a billiard ball and often exhibit chaotic behavior. In this chapter we begin with ballistic weak localization – a key concept of quantum transport in systems that show chaos in their classical dynamics. This phenomenon is an amusing quantum interference effect caused by the presence of time-reversal symmetry, and should not be confused with localization of the *kicked rotator* in momentum space intensively studied since the dawn of quantum chaos. Then we shall go on to more interesting themes related to quantum interference, namely, (1) ballistic Aharonov–Bohm oscillations in magneto-conductance, (2) quantitative evaluation of ballistic conductance fluctuations based on the extended semiclassical theory that includes the effect of electron diffraction, and finally (3) the fractal conductance fluctuations widely observed in experiments. In the last theme, the self-similar magneto-conductance fluctuations are pointed out to be caused by the self-similar unstable periodic orbits generated through a sequence of isochronous pitchfork bifurcations of straight-line librating orbits oscillating towards the harmonic saddles that are created universally as a consequence of softwall confinement.

7.1 Chaos and Quantum Transport

As we said in Chapter 1, an accumulation of experiments has reported the presence of anomalous fluctuations in the mesoscopic (nanoscale) world. These fluctuations are distinct from those arising from thermal noise or impurity potentials, but are taken as a quantal hallmark of chaos, i.e., quantum chaos. It is now becoming one of the big themes in mesoscopic physics to capture quantum chaos. In fact, by applying nanotechnology, quantum dots such as **stadium billiards** and antidots such as **Sinai billiards** have been fabricated at the interface of GaAs/AlGaAs semiconductor heterojunctions. In quantum dots, an electron is colliding with the inner side of concave billiard walls, whereas in antidots electrons collide with the outer side of convex billiard walls. The linear dimension of these systems is less than the electronic mean free path, and each electron there exhibits a deterministic motion like a ballistic missile except when it bounces elastically off the billiard wall. The activities concentrate on the measurement of quantum transport in these ballistic microstructures.

From the viewpoint of classical dynamics, the electronic motion in stadium or Sinai billiards is completely chaotic. Because the size of these billiards is typically of nanoscale, however, one should treat them within the framework of quantum mechanics and particularly calculate electrical conductance to compare with experimental results. As a result of this procedure, chaos in the microscopic world comes to manifest iteslf through the conductance. The conductive region at the interface of semiconductor heterojunctions has the following properties:

(1) By controlling the gate voltage, the electron concentration and thereby the Fermi energy E_F are tunable.
(2) At sufficiently low temperatures, the phase coherence length is larger than the system's size, making many-body effects negligible.

These characteristics, which are proper to systems with a few degrees of freedom, are very convenient to the investigation of the underlying nonlinear dynamics.

Now, consider a 2-d regular array of macroscopic hard disks with radius R (see Fig. 7.1), and let Ω be the area of the unit cell. This system, in which a point particle moves ballistically between disks and repeats collision with disk walls, is called the Sinai billiard system, giving an example of the simplest conservative dynamical system. The magnitude of the velocity v remains invariant against each elastic collision with disks.

We shall sketch the billiard-ball dynamics from the viewpoint of transport or electrical conductivity, while intensive insight was already made into these dynamics in Chapter 3. The number of bounces with disks per unit time is

$$\tau^{-1} = \frac{2Rv}{\Omega},\tag{7.1}$$

which is just the number of unit cells within the rectangular area $2Rv$ formed by an assembly of disks with diameter $2R$ moving with velocity v towards a rest ball. τ means the free flight time between successive collisions. A reference ball launched at

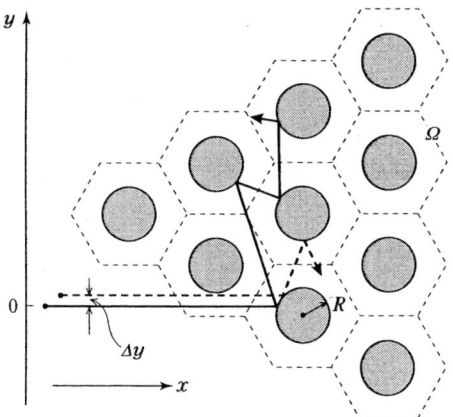

Fig. 7.1   Particle motion in Sinai billiards

an initial height ($y = 0$) horizontally (in the x direction) will collide with the first disk on the right and then repeats collisions one after another. Let us consider another ball with a small shift Δy at the initial height, which will move in a thoroughly different way after the first collision. Noting that the corresponding shift in the y component of the velocity vector just after the first collision is $\Delta v = \dfrac{v}{R}\Delta y$, the *magnified shift* $\Delta y^{(1)}$ in the y direction after the succeeding free flight (flight time τ) between the first and second collisions is given by

$$\Delta y^{(1)} = \Delta y + \tau \Delta v = \left(1 + \frac{v\tau}{R}\right)\Delta y, \tag{7.2}$$

and, at time $T = n\tau$ after the $n\,(\gg 1)$-th collision,

$$\Delta y^{(n)} = \left(1 + \frac{v\tau}{R}\right)^n \Delta y \sim \left[\exp\left(\frac{v}{R}T\right)\right]\Delta y, \tag{7.3}$$

implying an **exponential growth** of the initial small deviation. This simple argument tells us that Sinai billiards is a typically chaotic system with **Lyapunov exponent** $\lambda = \dfrac{v}{R}$. Then, the autocorrelation function for this system reads

$$\langle v(0)v(t)\rangle = \langle v(0)v(0)\rangle \exp(-\lambda t), \tag{7.4}$$

where $\langle \cdots \rangle$ denotes the average over initial velocities.

Let us turn to the Sinai billiard system on the submicron scale with the billiard ball replaced by a classical electron. Ignoring the electron–electron interaction, we shall move on to the calculation of electrical conductivity. With use of the electron density n_e, charge e and mass m_e, the electrical conductivity for a classical electron gas is given by the classical **Kubo formula**,

$$\sigma(\omega) = \frac{n_e e^2}{m_e} \int_0^\infty \frac{\langle v(0)v(t)\rangle}{\langle v(0)v(0)\rangle} \exp(i\omega t)\,dt. \tag{7.5}$$

A combination of (7.4) with (7.5) gives

$$\sigma(\omega) = \frac{n_e e^2}{\lambda m_e} \frac{1}{1 - i\dfrac{\omega}{\lambda}}\quad, \tag{7.6}$$

which means that the **Drude conductivity** is expressed in terms of a global characterisic quantity of chaos like the Lyapunov exponent, λ. The above scenario[78] suggests we take the conductivity as a measure of chaos, just as Abrahams *et al.*'s scaling theory of localization took the conductance as a measure of the randomness of the potential in dirty metals.

A right and proper treatment of quantum transport like electrical conductance, however, should be based on quantum mechanics. One of the most important phenomena in ballistic quantum transport would be the weak localization caused by quantum interference. So we shall elucidate how quantum chaos is related to ballistic weak localization or *vice versa*. In what follows, as in Chapter 4, we shall choose

open billiards with attached lead wires, and, within the framework of the semi-classical theory, investigate their conductance, revealing the quantum correction due to chaos. The quantum transport in antidot lattices like the lattice Sinai billiards discussed above will systematically be studied in Chapters 8 and 9.

7.2 Ballistic Weak Localization (WL)

What is weak localization? In a diffusive regime where the electronic mean free path is less than the device size, **Anderson localization** is the most fundamental concept in quantum transport. According to the idea of this localization, a strong random potential due to impurities makes an electron in the Bloch state extended throughout the whole system to localize with a finite extension of its wavefunction, leading to a suppression of the current. However, even in the metallic regime with Fermi wavelength less than the mean free path, there occurs a precursor of the localization called **weak localization**. Consider an orbit that comes back to the original position through scattering with impurities. In the absence of a magnetic field, one may expect a time-reversal partner of this back-scattering orbit. The quantum interference between a pair of time-reversal symmetric back-scattering orbits constitutes a standing wave, leading to the suppression of electronic transmission, namely, weak localization.

Since the early 1990s, a lot of experiments have come to be engaged in quantum transport in the ballistic regime, reporting **ballistic weak localization (WL)**, a phenomenon analogous to WL in the diffusive regime. Furthermore, this new phenomenon has turned out to be related to the nonintegrability feature, chaotic or regular, of the underlying classical electron dynamics. The first striking experiments on quantum transport in (classically) chaotic systems were carried out by Marcus *et al.*[37, 38] and by Chang[39, 40]. Figure 1.2 is the result by Marcus *et al.*, who measured the magnetic-field dependence of the conductance through billiards (quantum dots) with a pair of attached lead wires. As explained in Chapter 1, their experiments revealed the difference in the power spectra of conductance fluctuations between chaotic (stadium) and regular (circle) billiards. Besides this, the zero-field peak in the magneto-resistance is pointed out to be attributable to ballistic WL[39, 40], with the peak profile **Lorentzian** or **cusp-like** for the stadium or circle, respectively (see Chapter 1).

Ballistic WL is unambiguously explained as follows (Baranger *et al.*[46, 47]): Suppose an orbit is entering from the left lead to the cavity (dot) and, after bouncings with the cavity wall, returns back to the original entrance. In the zero field, there exists its time-reversal symmetric partner. Quantum interference between this pair of orbits forms a standing wave along the orbits, resulting in the enhancement of back-scattering, namely, the ballistic WL. The application of a weak magnetic field breaks the time-reversal symmetry, suppressing the WL effect and eventually reducing the magneto-resistance, which is the phenomenon of so-called **negative magneto-resistance**. The process of this reduction depends largely on the underlying electron dynamics and its statistical features whose prototype is the area distribution $P(\Theta)$.

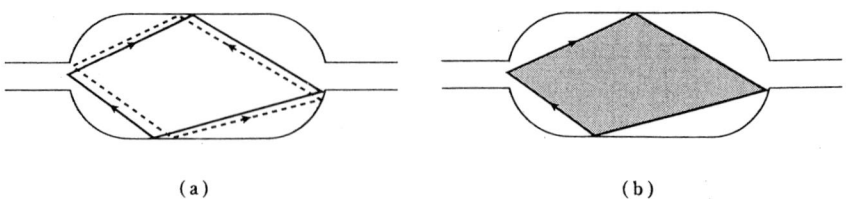

(a)   (b)

Fig. 7.2   (a) A pair of time-reversal symmetric back-scattering orbits; (b) area enclosed by a back-scattering orbit.

As described in Chapters 2 through 4, Θ is $2\pi\times$ the effective area enclosed by almost closed orbits formed by back-scattering orbits, and takes positive and negative values for clockwise and counter-clockwise rotations, respectively.

In the asymptotic region when $|\Theta| \gg \Theta_0$ with Θ_0 being the area of a given billiard system, $P(\Theta)$ obeys the exponential and power laws for chaotic and regular billiards, respectively:

$$P(\Theta) \propto \exp\left(-\alpha_{\text{cl}}\left|\Theta\right|\right), \tag{7.7a}$$

$$P(\Theta) \propto \frac{1}{\Theta^z} \tag{7.7b}$$

with the constant α_{cl} proper to individual billiards (e.g., the inverse of the area of a given billiard system). The above distributions indicate that orbits in a regular billiard system can easily form large areas thanks to orbital *stability*.

Noting that the magnetic field B is conjugate to the area Θ, we consider the Fourier transform of $P(\Theta)$, getting a quantum correction $\delta R_{\text{D}}(B)$ to the magneto-resistivity. For chaotic systems, we have

$$\delta R_{\text{D}}(B) = \frac{r_{\text{cl}}}{1 + \left(\dfrac{2B}{\alpha_{\text{cl}}\Phi_0}\right)^2}, \tag{7.8}$$

which is Lorentzian, whereas, for regular systems,

$$\delta R_{\text{D}}(B) \propto c_1 - c_2 \left|B\right|^{z-1}, \tag{7.9}$$

which is cusp-like (c_1, c_2 are positive constants). In (7.8), $\Phi_0 = \dfrac{hc}{e}$ is the flux quantum, and r_{cl} is the classical reflection probability defined as $r_{\text{cl}} = \dfrac{R_{\text{cl}}}{N_M}$ with R_{cl} the classical reflection coefficient and N_M the total mode number of the incident electron. The above results tell us how the difference in profiles of the weak-localization peaks reflects that of the area distributions. Beow, we shall derive the **weak-localization corrections** in (7.8) and (7.9) in more detail by having recourse to the semiclassical **Landauer formula**.

Consider a quantum billiard system (*characteristic length L*) attached with a pair of lead wires (*width W*) and let the range of modes $m(n)$ at the entrance lead be $1 \leq m, n \leq N_M$. Since in the present context we are concerned with the reflection coefficient, the exit is identified with the entrance. (By contrast, the transmission coefficient was treated in Chapter 4.) The reflection components of the scattering (S) matrix are

$$r_{n,m} = \delta_{n,m} - i\hbar\sqrt{v_n v_m} \int dy' \int dy \phi_n^*(y')\phi_m(y) \times G(0, y'; 0, y; E),$$

(7.10)

where $v_m(v_n)$ and $\phi_m(\phi_n)$ at each mode $m(n)$ are the longitudinal component (parallel to the lead) of the velocity vector and the transverse component of the wavefunction, respectively. G is the retarded Green function for an electron with Fermi energy $E = E_F = \dfrac{(\hbar k_F)^2}{2m_e}$ propagating from the initial (x, y) to final (x', y') points. In the limit $\hbar \to 0$, the semiclassical path-integral representation for G gives

$$G_{\rm scl}(y', y; E) = \frac{2\pi}{(2\pi i\hbar)^{3/2}} \sum_{s(y,y')} \sqrt{D_s} \exp\left(\frac{i}{\hbar} S_s(y', y; E) - i\frac{\pi}{2}\mu_s\right).$$

(7.11)

Here S_s is a (reduced) action for the classical path s of an electron with energy $E = E_F$, the Jacobian D_s arising from the quantum fluctuation around s represents the orbital stability, and the Maslov index μ_s is the number of conjugates points (singular points of D_s) along s. By substituting (7.11) into (7.10) and evaluating the double integral in the stationary-phase approximation, one obtains (see (4.52) and (4.53) in Chapter 4)

$$r_{n,m} = -\frac{\sqrt{2\pi i\hbar}}{2W} \sum_{s(\bar{n},\bar{m})} {\rm sgn}(\bar{n})\,{\rm sgn}(\bar{m})\sqrt{\tilde{D}_s} \exp\left(\frac{i}{\hbar}\tilde{S}_s(\bar{n}, \bar{m}; E) - i\frac{\pi}{2}\tilde{\mu}_s\right)$$

(7.12)

with $\tilde{D}_s, \tilde{S}_s, \tilde{\mu}_s$ evaluated at the saddle points. The $\delta_{n,m}$ term in (7.10) has been cancelled by the contribution from the zero-length orbit. On the right-hand side of (7.12), the summation is taken over the paths with incidence angle $\sin\theta = \dfrac{\bar{m}\pi}{k_F W}$ and ejection angle $\sin\theta' = \dfrac{\bar{n}\pi}{k_F W}$ at the initial (x) and final (x') cross-sections, respectively (Note: $\bar{m} = \pm m, \bar{n} = \pm n$).

(7.12) gives the reflection coefficient responsible for the magneto-resistance,

$$R = \sum_{n,m=1}^{N_M} |r_{n,m}|^2 \approx R_{\rm cl} + \delta R.$$

The quantum correction δR to the classical reflection coefficient $R_{\rm cl}$ consists of the diagonal $\delta R_{\rm D}$ and nondiagonal $\delta R_{\rm OD}$ components:

$$\delta R = \delta R_{\rm D} + \delta R_{\rm OD}$$

$$= \frac{1}{2} \frac{\pi}{k_{\mathrm{F}} W} \left[\sum_n \sum_{s \neq u} F_{n,n}^{s,u} + \sum_{n \neq m} \sum_{s \neq u} F_{n,m}^{s,u} \right] \tag{7.13}$$

with s, u representing classical paths satisfying the initial and final conditions mentioned below (7.12). By introducing for each path s the effective length (total flight length within a billiard system) $\widetilde{L}_s = \dfrac{\widetilde{S}_s}{k_{\mathrm{F}} \hbar}$ derived from the definition $S_s \equiv \displaystyle\int \boldsymbol{p}_s \cdot d\boldsymbol{l}$, we obtain an explicit form for $F_{n,m}^{s,u}$:

$$F_{n,m}^{s,u}(k_{\mathrm{F}}) = \sqrt{\widetilde{A}_s \widetilde{A}_u} \exp\left[ik_{\mathrm{F}}(\widetilde{L}_s - \widetilde{L}_u) - i\frac{\pi}{2}\mu_{s,u} \right], \tag{7.14}$$

where $\widetilde{A}_s = \left(\dfrac{\hbar k_{\mathrm{F}}}{W}\right) \widetilde{D}_s$ and $\mu_{s,u} \,(= \tilde{\mu}_s - \tilde{\mu}_u)$ is the difference of the Maslov indices between paths s, u.

In the following, we shall employ the diagonal approximation that retains only the diagonal term with respect to modes, δR_{D}, on the right-hand side of (7.12). Because the weak-localization correction is insensitive to Fermi wave number k_{F}, we proceed to average δR_{D} over all possible k_{F} under the many modes assumption $N_M \gg 1$. Through this procedure, only pairs of orbits satisfying $\widetilde{L}_s = \widetilde{L}_u$ can survive and contribute to the quantum interference term. In the absence of a magnetic field, the equality $\widetilde{L}_s = \widetilde{L}_u$ would be satisfied, provided u is a time-reversal partner of s. (The contribution from the term with $u = s$, as already included in R_{cl}, should not be incorporated into δR_{D}.) For the sake of brevity, the notation $\delta R, \delta R_{\mathrm{D}}$ is again used after their averaging over wave numbers.

On switching the magnetic field B, so long as it is sufficiently weak to guarantee almost straight orbits, the above argument should remain unchanged. However, there appears a net phase difference between a pair of mutually time-reversal symmetric orbits. Noting the action

$$S \equiv \int \boldsymbol{p} \cdot d\boldsymbol{l} = \int \left(m_e \boldsymbol{v} - \frac{e}{c} \boldsymbol{A} \right) \cdot d\boldsymbol{l}$$

with gauge potential $\boldsymbol{A}$, we envisage

$$\frac{1}{\hbar}(S_s - S_u) = -\frac{2e}{c\hbar} \int_s \boldsymbol{A} \cdot d\boldsymbol{l} = \frac{2\Theta_s B}{\Phi_0}.$$

Θ, as has often appeared so far, is $2\pi\times$ the effective area enclosed by almost closed **back-scattering orbits**. (Θ is chosen positive for the effectively clockwise rotation.) It is important in the above result that twice the contour integral is brought out by the interference between a pair of time-reversal symmetric orbits. The following procedure is parallel to Chapter 4, and we find (7.13) to be reduced to

$$\delta R_{\mathrm{D}}(B) \approx \int_{-\infty}^{\infty} d\Theta P(\Theta) \exp\left(\frac{2i\Theta B}{\Phi_0}\right).\tag{7.15}$$

(7.15) lets us understand why the area distributions $P(\Theta)$ in (7.7a) and (7.7b) lead to the weak-localization correction in (7.8) and (7.9) and why the weak-localization peak profile differs between stadium and circle billiards.

It has become crystal-clear: the ballistic WL is attributed to the quantum interference between a pair of time-reversal symmetric back-scattering orbits, and quantum chaos turns out to appear through quantum interference effects. In particular, we should emphasize that magneto-resistance and conductance are described only in terms of global properties of classical nonlinear dynamics.

7.3 Criticism against the Semiclassical Theory of Ballistic WL

Two important elements of the ballistic WL theory in the previous section are (i) it applies to the regime where the mean level spacing is much less than the energy broadening due to the opening of cavities, and (ii) an integration over all wave numbers is needed to obtain the lineshape. To satisfy (i), one should invent large dots with wide openings, which might however smear out the ballistic nature of quantum transport. To be consistent with (ii), an ensemble average (average over gate voltage) and/or thermal average for a single dot should be performed. However, in many cases, the dots are relatively small and measurements are made with only a few modes passing through the quantum point contacts (QPC's). Therefore it is not clear how much averaging is really sufficient. In fact, later experiments by Bird *et al.*[41] and numerical simulations by Akis *et al.*[82] showed several counterexamples. Various lineshapes for the resistance as a function of the magnetic field, including both linear and Lorentzian shapes, can be obtained in a fixed geometry (a single dot) as the gate voltage (energy) and/or the range of energy averaging (temperature) are varied. In these few mode cases, the profile of the zero-field resistance peak is determined by a conductance resonance structure that reflects the underlying energy spectrum of quantum dots. The semiclassical theory of ballistic WL can nontheless coexist with these interesting findings. The former is justified in the semiclassical (many-mode) limit, while the latter, in the quantal (a few mode) limit. As emphasized throughout the text, the semiclassical theory is the only way to capture the quantal manifestation of classical nonlinear dynamics.

The semiclassical theory of ballistic WL was originally developed by Baranger *et al.* Aside from the above practical criticism, this theory involves a serious problem: Although their qualitative derivation of the weak-localization peak profiles, (7.8) and (7.9), was very successful, quantitative evaluation of the magneto-resistance or conductance was not achieved. In particular, Baranger *et al.*'s theory ignored the nondiagonal component $\delta R_{\mathrm{OD}}(B)$ in the reflection coefficient. The quantal numerical calculation of their own, however, gives

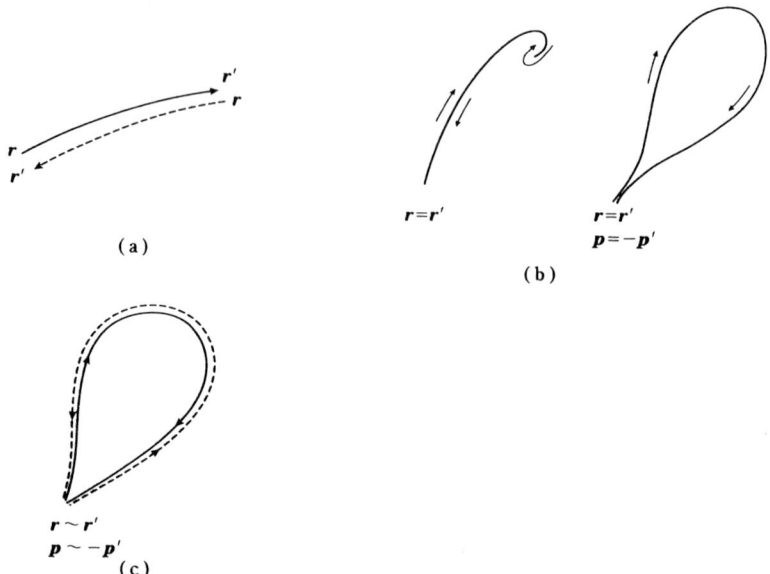

Fig. 7.3   A pair of time-reversal symmetric orbits: (a) a general pair; (b) self–retracing (U-turn) orbit (left) and back-scattering orbits in the limit $\hbar \to 0$ (right); (c) back-scattering orbits in case of $\hbar \neq 0$.

$$\frac{\delta R_{\mathrm{OD}}(B)}{\delta R_{\mathrm{D}}(B)} \approx -\frac{1}{2},$$

suggesting $\delta R_{\mathrm{OD}}(B)$ to be of the same order as $\delta R_{\mathrm{D}}(B)$. This fact is indicative of the unfavourable possibility that, with inclusion of $\delta R_{\mathrm{OD}}(B)$, the quantum correction might come to vanish!

Argaman[79], inferring that the semiclassical theory of the ballistic WL in quantum dots based on the Landauer formula should inevitably break the unitarity or the current conservation law, chose to apply the semiclassical theory to Kubo's linear response formula (to be described in Chapter 8) rather than to Landauer's. And, noting that a simple application of the stationary-phase approximation generates merely a self-retracing orbit going to and from a turning point, he concluded the absence of quantum correction. We shall explain his argument in more detail.

As already mentioned, the ballistic WL comes from the fact that, in the absence of the magnetic field, the interference among a pair of time-reversal symmetric back-scattering orbits forms a standing wave which suppresses the electron transmission. Let us start with the quantum correction to the conductivity (see Chapter 8) in the zero-field ($B = 0$) case. Suppose each classical orbit has initial momentum p and position r and final values p' and r'. Then, choose a pair of orbits γ_1 and γ_2 that are time-reversal symmetric with each other (see Fig. 7.3(a)). They have equal actions and Maslov indices and seem to bring about the quantum interference effect. Thanks to the time-reversal symmetry, this pair satisfies equality among the momenta as

$$p_2' = -p_1, \qquad p_1' = -p_2. \tag{7.16}$$

In the semiclassical path-integral expression of the Kubo formula for the conductivity (see Chapter 8), Argaman integrated the quantum correction over the final-point variables p', r' in the saddle-point approximation: the saddle point condition is now

$$p_1 - p_2 = 0, \qquad p_2' - p_1' = 0 \tag{7.17}$$

with $r' = r$ for each orbit. He obtained the explicit formula for the WL correction in the form of an average over the phase space:

$$\delta\sigma_{xx}^{\mathrm{wl}} \cong -\frac{e^2}{hm_e^2 V} \int_0^\infty dt \int d^2p\,d^2r\, \frac{l(p,r)}{v_{\mathrm{F}}} p_x p_x f(r, -p, t; r, p) \exp\left(-\frac{t}{\tau_{el}}\right), \tag{7.18}$$

where m_e, V, v_{F} and τ_{el} are the electron mass, area of the system, Fermi velocity and elastic scattering time, respectively. Within classical dynamics, the orbit satisfying both (7.16) and (7.17) is limited to that with $r' = r, p' = -p$, which is nothing but the self-retracing orbit going to and from a turning point (see the left of Fig. 7.3(b)). The self-retracing orbit is identical to its time-reversal partner: $\gamma_1 = \gamma_2$. Since this pair of trivial identical orbits was already incorporated in the classical conductivity, one should conceive the nontrivial pair of back-scattering orbits on the right of Fig. 7.3(b). In (7.18), $l(p,r)$ is the effective linear dimension of the range where these nontrivial closed orbits can exist. $f(r, -p, t; r, p)$ is the distribution function of such nontrivial orbits. The negative sign on the right-hand side of (7.18) naturally means the decrease of conductivity due to the ballistic weak localization.

The closed orbit on the right of Fig. 7.3(b), which has antiparallel momenta $p' = -p$ at the identical position $r' = r$, is anomalous and cannot indeed be generated within classical dynamics. To overcome this difficulty, however, Argaman pointed out the finiteness of the Planck constant, thereby accommodating the back-scattering orbits with their momenta at initial and final points being not exactly antiparallel with each other ($p' \sim -p$) (see Fig. 7.3(c)). These kinds of orbits, which have non-zero area, together with their time-reversal partners would cause the quantum interference effect, making possible a quantitative evaluation of the ballistic WL correction. His assertion inevitably leads to the conclusion that "the Kubo formulas + semiclassical theory" cannot derive the WL correction and that one must go beyond the lowest-order semiclassical approximation to quantum transport. This means that the inclusion of diffraction effects such as the small-angle-induced diffraction is essential, and this will be described in Section 7.5

On the other hand, Aleiner and Larkin[80] address the role of interference between long orbits with flight time longer than the Ehrenfest time when the classical–quantal correspondence should break down. Finally, Stone[81] declares that the ballistic WL has nothing to do with the semiclassical phenomena. To summarize all these criticisms, even though Kubo's linear response formula would have been employed instead of Landauer's, a right and proper theory of WL cannot be available, so long as one manages the semiclassical path-integral representation of Green functions and supplemental stationary-phase approximations.

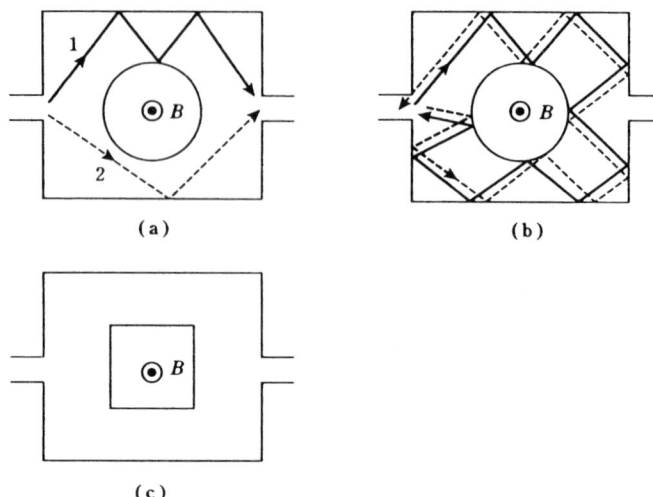

Fig. 7.4   Quantum interference effects: (a) interference between a pair of transmission orbits (single Sinai billiard system); (b) interference beween a pair of time-reversal symmetric back-scattering orbits (single Sinai billiard system); (c) double-squares billiards.

Existing theories on ballistic WL are far from being complete in both a quantitative evaluation and a way of thinking, and it is desirable to radically improve the semiclassical theory of quantum transport in chaotic systems. In the next section, we shall demonstrate recent progress in the semiclassical theory, which includes some remedies for the above difficulty. We shall first treat the Aharonov–Bohm effect, a theme much more interesting than ballistic WL.

7.4 Ballistic AAS Oscillation

Among the well-known quantum interference effects, the most interesting one is the **Aharonov–Bohm (AB) effect**. In this section, we shall investigate how chaos affects AB effects in ballistic systems (Kawabata and Nakamura[83]). Let us consider a square-shaped ballistic quantum dot (*characteristic length L*) with magnetic flux Φ penetrating through the circular hollow at the center (see Fig. 7.4) and investigate its quantum transport. An electron gas is confined to the ring-like region between the square wall and circular obstacle. This quantum dot is called the **Aharonov–Bohm (AB) billiard system**. In the AB billiards, the magneto-resistance is expected to display AB effects (AB oscillation and **Altshuler–Aronov–Spivak (AAS) oscillation**) whose behaviors reflect gobal properties of the underlying classical dynamics.

Now, what is AB oscillation? Suppose an electron is transmitting from the left lead to the right through the AB billiard system (see Fig. 7.4(a)). The wavefunction $\Psi(r; A)$ at the right exit is given as a sum of the one Ψ_1 through the path 1 and another Ψ_2 through the path 2, where

$$\Psi_i(r; A) = \exp\left[-i\frac{2\pi}{\Phi_0}\int_i A(l) \cdot dl\right] \Psi_i(r) \tag{7.19}$$

with $\Phi_0 = \dfrac{hc}{e}$ the flux quantum and $i = 1, 2$. ($\Psi_i(r)$ represents the wavefunction in the absence of AB flux.) It is essential in (7.19) that, despite the absence of a magnetic field exerting a ring-like conducting region, wavefunctions are affected by a gauge potential $A\,(B = \nabla \times A)$. Defining the AB flux as a contour integral along a ring-like region,

$$\Phi = \oint A \cdot dl, \tag{7.20}$$

the probability amplitude for an electron at the exit is given by

$$I = |\Psi_1(r, A) + \Psi_2(r, A)|^2$$
$$\cong 2\,|\Psi_1(r)|^2 \left\{1 + \cos\left(2\pi\frac{\Phi}{\Phi_0} + \eta\right)\right\}, \tag{7.21}$$

manifesting a Φ_0-periodic AB oscillation. The initial phase η in (7.21), originating from the difference of lengths between paths 1 and 2, depends on the Fermi energy E_F. In an electron gas at finite temperatures, E_F has finite width. Noting this fact, the averaging of (7.21) over E_F is inevitable, which eventually smears out the AB oscillation.

On the contrary, in the case of the interference that led to weak localization, namely, for the interference between a pair of time-reversal symmetric back-scattering orbits in Fig. 7.4(b), the oscillation term has exactly zero initial phase, and survives even after the averaging over E_F. As a consequence, the magneto-conductance shows $\dfrac{\Phi_0}{2}$-periodic AAS oscillation.

We now proceed to derive an explicit form for the AAS oscillation. An expression for the quantum correction in the reflection coefficient is the same as in the previous section: averaging of δR_D (see (7.13)) over wave number leaves behind a set of orbital pairs with $\tilde{L}_s = \tilde{L}_u$, which, in the absence of magnetic flux, is satisfied only when u is a time-reversal partner of s. (Note again that the contribution from the term with $u = s$, as already included in R_{cl}, should not be incorporated into δR_D.)

In the presence of the magnetic flux that penetrates only through the hollow of AB billiards, there appears a phase difference between a pair of mutually time-reversal symmetric orbits in Fig. 7.4(b). Assuming the smallness of the degree of opening $\left(\dfrac{W}{L}\right)$, back-scattering orbits are taken as almost closed ones, so the above phase difference is given by

$$\frac{1}{\hbar}(S_s - S_u) \approx \frac{2e}{ch} \oint_s \boldsymbol{A} \cdot d\boldsymbol{l} = 4\pi \frac{\Phi}{\Phi_0} w_s \qquad (7.22)$$

with Φ the AB flux and w_s the winding number around a central circle, e.g., around the flux. One should note that, **in contrast to the ordinary AB effect, the contour integral is doubled**. The quantum correction term can be obtained by applying a scheme analogous to (7.15). Here, let us first define a dwelling time T that an incident electron will spend in the cavity (quantum dot) until getting out to either one of the leads, and reorder orbits according to increasing T. Replacing the summation over s by integration over dwelling time, one gets an integral represention for the quantum correction term,

$$\delta R_{\mathrm{D}}(\Phi) \approx \int_{T_0}^{\infty} dT\, P(T) \sum_{w=-\infty}^{\infty} \mathcal{P}(w;T) \exp\left(4\pi i \frac{\Phi}{\Phi_0} w_s\right). \qquad (7.23)$$

Here $P(T)$ is the **dwelling time distribution** defined above T_0 (the time for the shortest orbit surrounding the inner circle), and $\mathcal{P}(w;T)$, the conditional probability, is the Gaussian winding number distribution with variance $\overline{w^2(T)} = \alpha \dfrac{T}{T_0}$ for an enssemby of orbits with a given T.

Similarly to the area distribution, classical dynamics for systems in Fig. 7.4 determines the dwelling time distribution in the AB billiard system. In the limit $\dfrac{T}{T_0} \gg 1$, the asymptotic behavior of $P(T)$ is

$$P(T) \sim \exp(-\gamma T), \qquad (7.24\mathrm{a})$$
$$P(T) \sim T^{-\beta} \qquad (7.24\mathrm{b})$$

for chaotic single Sinai billiards (Fig. 7.4(b)) and regular double-squares billiards (Fig. 7.4(c)), respectively. A power-law behavior like (7.24b) is also attributed to mixed billiards which accommodate both chaos and KAM tori (regular orbits) in their phase space.

In the case of a chaotic system (Sinai billiards), a combination of (7.23) with the exponential law (7.24a) leads to an explicit form for the weak-localization correction $\delta R_{\mathrm{D}}(\Phi)$[83],

$$\delta R_{\mathrm{D}}(\Phi) \sim \frac{1}{2}\sqrt{\frac{T_0}{2\gamma\alpha}} \frac{\sinh\left(\sqrt{\frac{2T_0\gamma}{\alpha}}\right)}{\cosh\left(\sqrt{\frac{2T_0\gamma}{\alpha}}\right) - \cos\left(4\pi\frac{\Phi}{\Phi_0}\right)}$$

$$= \frac{1}{2}\sqrt{\frac{T_0}{2\gamma\alpha}} \left\{ 1 + \sum_{n=1}^{\infty} \exp\left(-n\sqrt{\frac{2T_0\gamma}{\alpha}}\right) \cos\left(4\pi n\frac{\Phi}{\Phi_0}\right) \right\},$$

$$\qquad (7.25)$$

which manifests the ballistic AAS ($\dfrac{\Phi_0}{2}$-periodic) oscillation.

Surprisingly, this formula is quite similar to the quantum correction to Drude conductivity for a quasi-one-dimensional metallic ring in a diffusive regime, i.e., the diffusive AAS oscillation,

$$\Delta\sigma(\Phi) = \frac{e^2}{\pi^2 h} L_\varepsilon \frac{\sinh\left(\dfrac{L_y}{L_\varepsilon}\right)}{\cosh\left(\dfrac{L_y}{L_\varepsilon}\right) - \cos\left(4\pi\dfrac{\Phi}{\Phi_0}\right)} \tag{7.26}$$

with L_y the circumference of the ring and $L_\varepsilon = \sqrt{D\tau_\varepsilon}$ the phase coherence length expressed in terms of the diffusion constant D and phase relaxation time τ_ε. (7.26) was attained by applying perturbation theory to the Kubo formula, whereas (7.25) is the result of application of semiclassical scattering theory to Landauer's formula. The phase relaxation time (τ_ε) in (7.26) obviously corresponds to the characteristic dwelling time $\left(\dfrac{1}{\gamma}\right)$ in (7.25). However, the mystery is left unsolved as to why thoroughly different approaches have led to identical results.

By the way, what is the corresponding expression for the quantum correction in the case of regular (e.g., double squares) and mixed (e.g., double circles with unequal centers) billiards? The dwelling time distribution here obeys (7.24b), and therefore the functional form for δR_{D} more or less differs from (7.25). Qualitatively, however, we find litte essential diffference: irrespective of the integrability of AB billiards, chaotic or regular (and mixed), the primary $(n = 1)$ harmonics give a dominant $\dfrac{\Phi_0}{2}$-periodic term in δR_{D}. The amplitudes of the higher harmonics $(n \geq 2)$ decrease exponentially and algebraically with increasing n, respectively, in the chaotic and regular (and mixed) systems.

In real experiments where the magnetic field is exerted over not only the hollow but also the whole part of a billiard system, we can observe both (1) the ballistic weak localization peak or the so-called negative magneto-resistance and (2) the ballistic AAS oscillation with its amplitude decaying with the magnetic field[83].

7.5 Effects of Small-Angle Induced Diffraction

We have so far argued about quantum interference effects like ballistic weak localization (WL) and AAS oscillation within the framework of the semiclassical theory of quantum transport. Nevertheless, the correct quantitative evaluation of such effects has not yet been done, and demands an improvement of the semiclassical theory for the following reasons:

(i) The semiclassical theory cannot evaluate the nondiagonal component $\delta R_{\mathrm{OD}}(B)$ in the reflection coefficient.

(ii) This theory also fails to bring out the WL correction δT in the transmission coefficient T. More precisely, the quantum correction in the transmission coefficient should arise from interference among classical orbits coming from the entrance and getting out to the exit (different from the entrance!), but these orbits cannot be time-reversal symmetric with each other.

(iii) As a consequence of (i) and (ii), transmission and reflection coefficients cannot satisfy the unitarity condition, bringing about $\delta R + \delta T \neq 0$.

These facts indicate a fatal shortcoming of the semiclassical theory of quantum transport. To overcome such a problem, Takane and Nakamura proposed an **extended semiclassical theory** that accommodates the effect of small-angle induced diffraction[84, 85]. The essence of their theory will be given below.

7.5.1   Partial Time-Reversal Symmetry and Ballistic Weak-Localization Correction

Suppose a fully chaotic ballistic billiard system with a pair of attached leads exists, and consider the transmission of an electron with Fermi wavelength λ_F much less than the widths of the left (entrance) and right (exit) lead wires ($\lambda_F \ll W_L, W_R$). Let $T(R)$ be the transmission (reflection) coefficient and $N_{L(R)}$ the total mode number in the left (right) lead. ($N_R = N_L$ is not always necessary.) The unitarity condition is now given by $T + R = N_L$. T is related to the conductance G via the Landauer formula,

$$T = \frac{h}{2e^2} G.$$

Now, choosing the **real-space representation** instead of the mode representation used so far, let us write the transmission coefficient T in terms of the Kubo nonlocal conductivity as[84]

$$T = \frac{\hbar^4}{4m_e^2} \int_{C_R} dr_{1\perp} \int_{C_L} dr_{2\perp} D_{1,4} D_{2,3} K\left(\boldsymbol{r}_1, \boldsymbol{r}_2, \boldsymbol{r}_3, \boldsymbol{r}_4\right) \Big|_{r_4 = r_1, r_3 = r_2} \tag{7.27}$$

with C_R and C_L representing cross-sections of the right and left lead wires, respectively. Here the differential

$$D_{i,j} = \left(\frac{\partial}{\partial r_{i//}} - \frac{\partial}{\partial r_{j//}} \right) \tag{7.28}$$

stands for the momentum component parallel to the leads, and the **pair-propagator** K is defined by

$$K(\boldsymbol{r}_1, \boldsymbol{r}_2, \boldsymbol{r}_3, \boldsymbol{r}_4) = G^+(\boldsymbol{r}_1, \boldsymbol{r}_2; E_{\mathrm{F}}) G^-(\boldsymbol{r}_3, \boldsymbol{r}_4; E_{\mathrm{F}}) \tag{7.29}$$

with retarded (advanced) Green functions $G^+ (G^-)$ for an electron with Fermi energy E_{F} propagating from the entrance to the exit. It should be noted that (7.27) describes the trace of the absolute square of transmission components in the S matrix.

Let us evaluate T within the framework of the semiclassical theory. Substituting into (7.29) the semiclassical Green function

$$G^+(\boldsymbol{r}_1, \boldsymbol{r}_2; E_{\mathrm{F}}) = \frac{2\pi}{(2\pi i \hbar)^{3/2}}$$

$$\times \sum_s A_s \exp\left\{\frac{i}{\hbar}S_s(\boldsymbol{r}_1, \boldsymbol{r}_2; E_{\mathrm{F}}) - i\frac{\pi}{2}\mu_s\right\} \qquad (7.30)$$

with a classical path s, action S_s and Maslov index μ_s, we have from (7.27),

$$T = \frac{1}{8\pi\hbar m_e^2} \int_{C_R} dr_{1\perp} \int_{C_L} dr_{2\perp} \sum_{s,u}(p_{s//} + p_{u//})(p'_{s//} + p'_{u//})A_s A_u$$

$$\times \exp\left\{\frac{i}{\hbar}(S_s(\boldsymbol{r}_1, \boldsymbol{r}_2; E_{\mathrm{F}}) - S_u(\boldsymbol{r}_1, \boldsymbol{r}_2; E_{\mathrm{F}})) - \frac{i\pi}{2}(\mu_s - \mu_u)\right\},$$

$$(7.31)$$

where

$$\boldsymbol{p}_s = \frac{\partial S_s(\boldsymbol{r}_1, \boldsymbol{r}_2)}{\partial \boldsymbol{r}_1}, \qquad \boldsymbol{p}'_s = -\frac{\partial S_s(\boldsymbol{r}_1, \boldsymbol{r}_2)}{\partial \boldsymbol{r}_2}.$$

The classical transmission coefficient is obtained from the diagonal term $s = u$ in the double summations in (7.31). On the contrary, the WL correction arises from the interference between a pair of orbits with $s \neq u$. Noting both s and u are orbits starting at the left lead and ending at the right, however, there exists no time-reversal symmetric pair in the double sums. Consequently, as long as one continues to cling to the existing formalism of the semiclassical theory, it is not possible to describe the WL effects at all.

Consider a pair of classical (normal) orbits crossing each other with a very small intersection angle (see Fig. 7.5(a)) that will contribute to the semiclassical Green function. In treating chaotic systems, one would often encounter such a circumstance. By loosening the semiclassical constraint or, in other words, by going beyond the lowest-order semclassical approximation, one may consider another **pair of anomalous orbits** with each changing from one to another near the crossing point (see Fig. 7.5(b)). These anomalous orbits would naturally be generated via small-angle induced diffraction. The extended semiclassical theory which includes small-angle induced anomalous (diffractive) orbits makes it possible for us to conceive of a **pair of partially time-reversal symmetric transmission orbits** in Fig. 7.5(b), which are running antiparallel to each other in the intermediate region although parallel in the major part of the orbits. This pair of orbits, being mutually time-reversal symmetric in the intermediate region, are expected to contribute to the WL effect. This idea was originally conceived by Aleiner and Larkin.

In fact, for open chaotic billiards we can obtain a quantitatively correct quantum correction δT due to the interference between such pairs, by using a semiclassical **pair-propagator**. The result is (Takane and Nakamura[84])

$$\delta T = -t_{\mathrm{cl}}r_{\mathrm{cl}}, \qquad (7.32)$$

where $t_{\mathrm{cl}}(\equiv T_{\mathrm{cl}}/N_{\mathrm{L}})$ and $r_{\mathrm{cl}}(\equiv R_{\mathrm{cl}}/N_{\mathrm{L}})$ are classical transmission and reflection probabilities, respectively, and depend only on the size of the right and left holes, i.e., the widths of the lead wires. (7.32) implies a reduction in the transmission coefficient

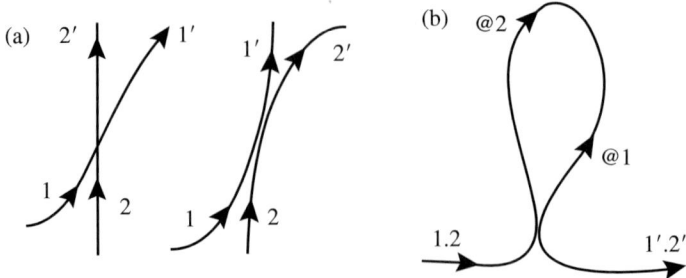

Fig. 7.5   (a) A pair of normal orbits (left) and that of small-angle induced anomalous (diffractive) orbits (right); (b) a pair of partially time-reversal symmetric transmission orbits that run antiparallel to each other in the intermediate region.

brought out by the ballistic WL effect. In other words, the interference between a pair of partially time-reversal symmetric back-scattering orbits have really suppressed the electronic transmission.

The WL correction to the reflection coefficient is composed of the contribution due to a pair of **partially time-reversal symmetric** back-scattering orbits $\delta R^{(1)}$ and that due to a pair of **totally time-reversal symmetric** back-scattering orbits $\delta R^{(2)}$. The former is $\delta R^{(1)} = -r_{\mathrm{cl}}^2$, decreasing the reflection, whereas $\delta R^{(2)} = r_{\mathrm{cl}}$. With use of probability conservation $t_{\mathrm{cl}} + r_{\mathrm{cl}} = 1$, the sum of WL corrections to the reflection coefficient now becomes

$$\delta R = \delta R^{(1)} + \delta R^{(2)} = t_{\mathrm{cl}} r_{\mathrm{cl}}. \qquad (7.33)$$

$\delta R^{(2)}$ agrees with Baranger et al.'s diagonal term δR_{D}, and $\delta R^{(1)}$ corresponds to the nondiagonal term δR_{OD} not derived so far. Remarkably, the ratio $\delta R_{\mathrm{OD}}/\delta R_{\mathrm{D}} = -r_{\mathrm{cl}}$ is almost identical to Baranger et al.'s numerical quantal calculation. Equations (7.32) and (7.33) ensure that the unitarity condition

$$\delta R + \delta T = 0,$$

together with the classical equality $T_{\mathrm{cl}} + R_{\mathrm{cl}} = N_L$, is satisfied. We come to the conclusion that the extended semiclassical theory that accommodates small-angle induced diffractions is crucial for us to evaluate quantitatively correct quantum interference effects and also to establish the theory satisfying unitarity. We also find the full expression for the WL correction in the presence of the B field. With use of (7.33), the formula (7.8) can be improved as[84]

$$\delta R(B) = \frac{t_{\mathrm{cl}} r_{\mathrm{cl}}}{1 + \left(\dfrac{2B}{\alpha_{\mathrm{cl}} \Phi_0}\right)^2} \ .$$

The recent semiclassical theory[86] of chaotic quantum transport attacked the same subject as above, and proposed an off-diagonal trajectory pair formed by a self-intersecting classical orbit and a neighboring orbit differing mainly in the region around the self-intersection. This pair, which appears in systems with uniformly hyperbolic dynamics, is exactly the same as that which is intensively explained in the present subsection.

7.5.2 Semiclassical Derivation of Universal Conductance Fluctuations

By developing the study in Chapter 4 in the direction of the extended semiclassical theory, Takane and Nakamura[85] also succeeded in evaluating **quantitatively correct ballistic conductance fluctuations**. In what follows, their theory will be explained. The conductance G and transmission coefficient T are related by the Landauer formula, $G = \dfrac{e^2}{\pi} T$. (In this subsection we employ the abbreviation $\hbar = 1$.) Again we start with (7.27).

The variance of T is

$$\mathrm{var}(T) = \langle (T - \langle T \rangle_0)^2 \rangle_0$$

with $\langle \cdots \rangle_0$ the average over E_{F}. A more explicit form is

$$
\begin{aligned}
\mathrm{var}(T) = \frac{1}{(4m_e^2)^2} &\int_{C_R} dr_{1\perp} \int_{C_L} dr_{2\perp} \int_{C_R} dr_{5\perp} \int_{C_L} dr_{6\perp} D_{1,4} D_{2,3} D_{5,8} D_{6,7} \\
&\times \left. \delta M(\boldsymbol{r}_1, \boldsymbol{r}_2, \boldsymbol{r}_3, \boldsymbol{r}_4; \boldsymbol{r}_5, \boldsymbol{r}_6, \boldsymbol{r}_7, \boldsymbol{r}_8) \right|_{r_4 = r_1, r_3 = r_2, r_8 = r_5, r_7 = r_6},
\end{aligned}
$$

$$(7.34)$$

where δM is the phase-space average of the quantum-interference correction to the product of pair-propagators

$$M = K(\boldsymbol{r}_1, \boldsymbol{r}_2, \boldsymbol{r}_3, \boldsymbol{r}_4) K(\boldsymbol{r}_5, \boldsymbol{r}_6, \boldsymbol{r}_7, \boldsymbol{r}_8).$$

(The average over E_{F} is assumed equivalent to that over the phase space.)

We shall move on to the Fourier transform of the pair-propagator. First, let us introduce one Fourier transform K^D as follows:

$$
\begin{aligned}
K(\boldsymbol{r}_1, &\boldsymbol{r}_2, \boldsymbol{r}_3, \boldsymbol{r}_4) \\
&= \int \frac{d\boldsymbol{p}_1 \, d\boldsymbol{p}_2}{(2\pi)^4} e^{i\boldsymbol{p}_1(\boldsymbol{r}_1 - \boldsymbol{r}_4) - i\boldsymbol{p}_2(\boldsymbol{r}_2 - \boldsymbol{r}_3)} K^D(\boldsymbol{p}_1, \boldsymbol{R}_1; \boldsymbol{p}_2, \boldsymbol{R}_2)
\end{aligned}
$$

$$(7.35)$$

with the center-of-mass coordinates

$$\boldsymbol{R}_1 = \frac{1}{2}(\boldsymbol{r}_1 + \boldsymbol{r}_4), \qquad \boldsymbol{R}_2 = \frac{1}{2}(\boldsymbol{r}_2 + \boldsymbol{r}_3).$$

K^D stands for the diffuson-type pair-propagator that describes the phase interference between an electron and a hole. Second, by means of other center-of-mass coordinates

$$\boldsymbol{R}_1 = \frac{1}{2}(\boldsymbol{r}_1 + \boldsymbol{r}_3), \qquad \boldsymbol{R}_2 = \frac{1}{2}(\boldsymbol{r}_2 + \boldsymbol{r}_4),$$

the Cooperon-type pair-propagator that describes the phase interference among electrons is given by

$$
\begin{aligned}
K(&\boldsymbol{r}_1, \boldsymbol{r}_2, \boldsymbol{r}_3, \boldsymbol{r}_4) \\
&= \int \frac{d\boldsymbol{p}_1 d\boldsymbol{p}_2}{(2\pi)^4} e^{i\boldsymbol{p}_1(\boldsymbol{r}_1 - \boldsymbol{r}_3) - i\boldsymbol{p}_2(\boldsymbol{r}_2 - \boldsymbol{r}_4)} K^C(\boldsymbol{p}_1, \boldsymbol{R}_1; \boldsymbol{p}_2, \boldsymbol{R}_2).
\end{aligned}
\tag{7.36}
$$

The above terminology of Cooperon is introduced in analogy to the Cooper pair in superconductivity that consists of electrons with mutually antiparallel wave vectors. Noting the energy conservation law, the Fourier transforms $K^F (F = D, C)$ can be rewritten explicitly in the semiclassical approximation as

$$
\begin{aligned}
K^F(&\boldsymbol{p}_1, \boldsymbol{R}_1; \boldsymbol{p}_2, \boldsymbol{R}_2) \\
&= \frac{2\pi}{g(E_\mathrm{F})} \delta\left(E_\mathrm{F} - \frac{p_1^2}{2m}\right) \delta\left(E_\mathrm{F} - \frac{p_2^2}{2m}\right) F(\boldsymbol{n}_1, \boldsymbol{R}_1; \boldsymbol{n}_2, \boldsymbol{R}_2)
\end{aligned}
$$

$$\tag{7.37}$$

with $g(E_\mathrm{F}) = \dfrac{m_e}{2\pi}$ the 2-d density of states and $\boldsymbol{n}_i = \dfrac{\boldsymbol{p}_i}{p_i}$ the unit vector parallel to the momentum. Below, we shall employ the abbreviation $1 \equiv (\boldsymbol{n}_1, \boldsymbol{R}_1)$, $2 \equiv (\boldsymbol{n}_2, \boldsymbol{R}_2)$, etc., and call the reduced pair-propagators $F = D$ and C in (7.37) simply diffuson and Cooperon, respectively. The reduced pair-propagators, which depend on the initial and final momenta and positions, behave as probability functions obeying the Liouville equation for the classical electron motion and satisfying the hard-wall boundary condition at billiard walls. As shown below, their solutions can be found through a phenomenological insight.

In the semiclassical limit, fine scales less than λ_F can be coarse-grained, and the pair-propagators after this smoothing are denoted as $\overline{D}(1,2), \overline{C}(1,2)$. Furthermore, thanks to the ergodicity in chaotic billiards, $\overline{D}(1,2)$ and $\overline{C}(1,2)$ are assumed to depend on neither the initial nor final momenta and positions. Since the diffusion and Cooperon satisfy the particle-number conservation law (an electron coming into a billiard system eventually gets out through either one of the right and left exits), we find

$$\overline{D}(1,2) = \overline{C}(1,2) \sim \frac{\tau_B}{\Theta_0} \tag{7.38}$$

where Θ_0 and τ_B^{-1} are the area of the billiard system and the escape rate of the electron, respectively. Let the partial escape rate $\tau_{L(R)}^{-1}$ through the right (left) lead be defined by

$$\frac{1}{\tau_{L(R)}} = \frac{1}{\Theta_0} \int_{C_{L(R)}} dR_\perp \int \frac{d\phi}{2\pi} H(\boldsymbol{n}_{/\!/}^{L(R)} \cdot \boldsymbol{n}) v_\mathrm{F} \cos\phi$$

$$= \frac{v_{\mathrm{F}} W_{L(R)}}{\pi \Theta_0},$$ (7.39)

and then the escape rate through both leads is

$$\tau_B^{-1} = \tau_L^{-1} + \tau_R^{-1}.$$

In (7.39), $n_{//}^{L(R)}$ is the unit vector outwardly normal to the cross-section of each lead $C_{L(R)}$, ϕ is the angle between $n_{//}^{R}(-n_{//}^{L})$ and n, and finally $H(\cdot)$ represents the Heaviside step function. The classical transmission (t_{cl}) and reflection (r_{cl}) probabilities are determined in terms of the partial escape rates:

$$t_{\mathrm{cl}} = \frac{\tau_B}{\tau_R},$$ (7.40a)

$$r_{\mathrm{cl}} = \frac{\tau_B}{\tau_L}.$$ (7.40b)

Specifically, in the case of $W_L = W_R$ (equal wire widths), we find $t_{\mathrm{cl}} = r_{\mathrm{cl}} = \frac{1}{2}$.

Within the framework of the extended semiclassical theory that incorporates small-angle induced diffractions, there are five kinds of combinations of paths that contribute to $\mathrm{var}(T)$ (see Fig. 7.6). Each of the four kinds of combinations except for the case (c) includes anomalous orbits induced by small-angle diffractions. Provided that one is concerned with purely classical orbits only, the case (c) alone survives. For details, readers should refer to the original paper[85]. Here we shall summarize the result. The process (a) yields

$$\mathrm{var}(T)_{\mathrm{a}} = \frac{\tau_B^4}{\tau_L^2 \, \tau_R^2} \int d_5 d_6 (-\hat{L}_5 \hat{L}_6) \left(\overline{D}(5,6)\right)^2.$$ (7.41)

Here $d_j \equiv \dfrac{d\boldsymbol{R}_j d\phi_j}{2\pi}$ with ϕ_j the angle of n_j and the integration over $\boldsymbol{R}_j$ being carried out inside a billiard system. $\hat{L}_j \equiv v_{\mathrm{F}} n_j \cdot \left(\dfrac{\partial}{\partial \boldsymbol{R}_j}\right)$ is the Liouville operator for classical electronic motion ("$\cdot$" denots the inner product). The total-differential feature of $\hat{L}_j$ facilitates the evaluation of (7.41), which now reads

$$\mathrm{var}(T)_{\mathrm{a}} = \frac{\tau_B^4}{\tau_L^2 \, \tau_R^2} \sum_{\alpha,\beta=L,R} \int_{C_\alpha} \frac{dR_{5\perp} d\phi_5}{2\pi} \int_{C_\beta} \frac{dR_{6\perp} d\phi_6}{2\pi}$$
$$\times H(\boldsymbol{n}_{//}^{\alpha} \cdot \boldsymbol{n}_5) H(-\boldsymbol{n}_{//}^{\beta} \cdot \boldsymbol{n}_6) v_{\mathrm{F}}^2 \cos\phi_5 \cos\phi_6 \left(\overline{D}(5,6)\right)^2.$$ (7.42)

With use of the definition for escape rate (7.39) together with

$$\left(\overline{D}(5,6)\right)^2 \sim \left(\frac{\tau_B}{\Theta_0}\right)^2$$

available from (7.38), the expression in (7.42) is reduced to

(a)

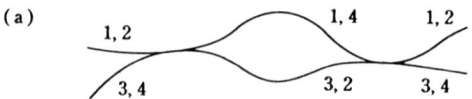

(b)

(c)

(d)

(e)

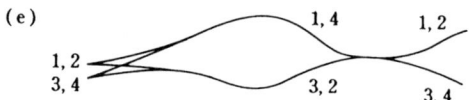

Fig. 7.6   Five kinds of processes contributing to var(T). Each process consists of four electronic paths transmitting from left to right. Paths 1, 2, 3, 4 correspond to propagators $G^+(\boldsymbol{r}_1, \boldsymbol{r}_2), G^-(\boldsymbol{r}_3, \boldsymbol{r}_4), G^+(\boldsymbol{r}_5, \boldsymbol{r}_6), G^-(\boldsymbol{r}_7, \boldsymbol{r}_8)$, respectively. Crossings among paths are due to the small-angle induced diffraction. Two pair-propagators (1, 4) and (3, 2) in the process (b) are Cooperons (reproduced from[85]).

$$\text{var}(T)_\text{a} = \frac{\tau_B^4}{\tau_L^2 \, \tau_R^2} = (t_\text{cl} r_\text{cl})^2. \tag{7.43}$$

Similarly, the contribution from the process (b) is available by replacing the integrand in (7.41) by $(\widehat{L}_5 \widehat{L}_6)\left(\overline{C}(5, 6)\right)^2$, and the result is

$$\text{var}(T)_\text{b} = \frac{\tau_B^4}{\tau_L^2 \, \tau_R^2} = (t_\text{cl} r_\text{cl})^2. \tag{7.44}$$

Further, we find the remaining contributions from the processes (c)–(e),

$$\text{var}(T)_c = t_{cl}r_{cl}, \qquad \text{var}(T)_d = -t_{cl}r_{cl}^2, \qquad \text{var}(T)_e = -t_{cl}^2 r_{cl}.$$

Because of probability conservation $(t_{cl} + r_{cl} = 1)$, however, these contributions come to vanish as a whole, namely,

$$\text{var}(T)_c + \text{var}(T)_d + \text{var}(T)_e = 0. \tag{7.45}$$

In conclusion, only the processes (a) and (b) prove to contribute to the conductance fluctuations, which now read

$$\text{var}(G) = \frac{2e^2}{\pi}(t_{cl}r_{cl})^2. \tag{7.46}$$

In particular, in the case $W_L = W_R$, (7.46) reduces to

$$\text{var}(G) = \frac{e^2}{8\pi}$$

in agreement with the prediction of random matrix theory. In the presence of a magnetic field that breaks the time-reversal symmetry, the contribution from the Cooperon process (b) vanishes and only the process (a) gives conductance fluctuations:

$$\text{var}(G) = \frac{e^2}{\pi}(t_{cl}r_{cl})^2, \tag{7.47}$$

which, in the case $W_L = W_R$, again reproduces the result of the random matrix theory,

$$\text{var}(G) = \frac{e^2}{16\pi}.$$

Thus, the extended semiclassical theory that includes anomalous orbits induced by small-angle diffractions can resolve the problem (: the breakdown of unitarity condition) involved in Baranger *et al.*'s weak-localization theory and also establish a quantitatively correct theory of ballistic conductance fluctuations.

As we have seen heretofore in this chapter, the semiclassical theory and its extended version make it possible for us to describe principal quantum interference effects in ballistic systems like weak localization, AAS oscillations and conductance fluctuations on the basis of classical pictures. In particular, it is amusing that the magneto-resistance and conductance are expressed in terms of the statistical properties (e.g., area distribution, dwelling-time distribution, and winding-number distribution) of nonlinear dynamics and chaos.

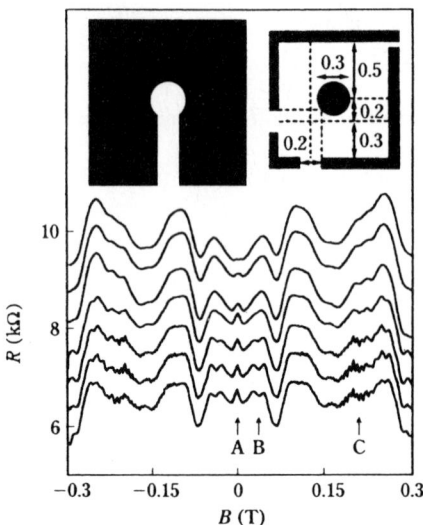

Fig. 7.7 Measured magneto-resistance. From the bottom curve, temperature grows as $30\,\mathrm{mK}, 0.1\,\mathrm{K}, 0.4\,\mathrm{K}, 0.8\,\mathrm{K}, 1.6\,\mathrm{K}, 2.5\,\mathrm{K}, 3.3\,\mathrm{K}$. The pair of upper insets indicates a single Sinai billiard system as an object of measurement (reproduced from[88]).

7.6 Self-Similar Magneto-Conductance Fluctuations

The last theme in this chapter is the self-similar magneto-conductance fluctuations observed widely in submicron-scale phase coherent ballistic billiards. There are abundant experiments on self-similar conductance fluctuations. Taylor *et al.*[88] fabricated a square billiard system with a side length of $1\,\mu$m at the interface of a high-mobility AlGaAs/GaAs heterojunction. A pair of lead wires with width of $0.2\,\mu$m are attached to the billiard system. They measured magneto-resistance (see Fig. 7.7) by applying the magnetic field B perpendicular to the system plane. Here the electronic mean free path is $25\,\mu$m which is less than the system size, and the Fermi wavelength is $0.05\,\mu$m (for an electron concentration of $2.3 \times 10^{11}\mathrm{cm}^{-2}$). In these circumstances, an electron can execute ballistic motion and the semiclassical treatment is justified. To realize a single Sinai billiard system, Taylor *et al.* applied a gate voltage at the center of the square, making a circular antidot (obstacle). The radius of the circle is tunable by varying the gate voltage. Figure 7.7 is the measured magneto-resistance for a single Sinai billiard system with a fixed aspect ratio (the radius of circle/side length of square).

We find in Fig. 7.7, besides the Shubnikov–de Haas oscillation (arrow C) in the B field above 0.15 T (tesla), a new quantum oscillation (arrow B) in the field below 0.05 T as well as the weak-localization peak (arrow A) in the vicinity of $B = 0$. The weak-localization peak is due to quantum interference between a pair of time-reversal

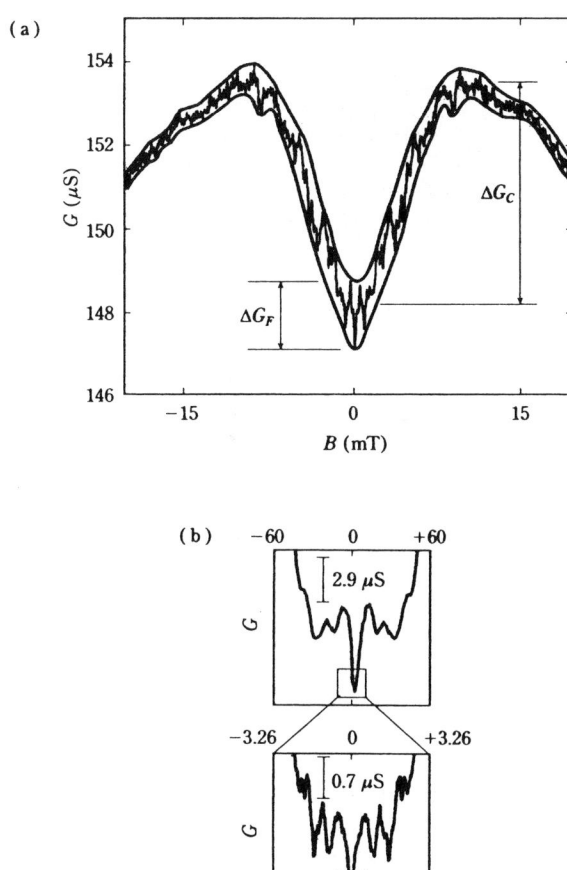

Fig. 7.8 (a) Coarse-grained structure for magneto-conductance G and small fluctuations; (b) fractal structure of G (reproduced from[88]).

symmetric back-scattering orbits. It is not easy, however, to determine whether the new oscillation should be attributed to AB oscillation or the aforementioned AAS oscillation. This difficulty is caused by the ambiguity of defining the magnetic flux in the present system where the B field is exerted over not only the circular hollow but also the whole part of the billiard system.

Another more important discovery in Taylor *et al.*'s experiment is the small fluctuations superimposed on the quantum oscillations (see Fig. 7.8(a)). In the conductance in the weak-field regime less than 0.05 T, they observed small-amplitude fluctuations on a (transverse) scale of 0.5mT superimposed on large-amplitude ones on a scale of 10 mT, and pointed out the presence of a **fractal structure** in the conductance: magnification of part of the structure reproduces a structure identical to the original one (see Fig. 7.8(b)). The larger the radius of the central circle, the more crystal-clear the fractal structure becomes. What is occurring in the underlying classical dynamics? On switching the weak magnetic field, the phase space structure for the Sinai billiard system and even for a simple square billiard system becomes mixed of chaos and tori. This tendency becomes more remarkable when billiard system's hard walls are replaced by soft walls[89-91]. In phase space, tori are embedded in the chaotic sea with each torus being accompanied by several small-scale islands. The magnification of each island reproduces the same torus with island structures as the original one. One is therefore inclined to interpret the experimental result as a quantum signature of the hierarchy of island structures in the mixed system.

In Chapters 1–4, we understood that the shape of the weak-localization peak in magneto-resistance near zero magnetic field is determined by the underlying classical dynamics, i.e., chaotic or regular, and the semiclassical theory explains this phenomenon very well. However, in generic classical systems, the dynamics is neither chaotic nor regular, but mixed of both. One of the most intriguing universal phenomena in such a case is the existence of a **hierarchical island structure** in its phase space. This self-similar structure is believed to make the escape time probability of the electron different from that from the fully chaotic cavity. We have a power-law distribution of escape time in general mixed systems, whereas an exponential law in fully chaotic systems. The success in semiclassical explanations of phase coherent ballistic phenomena naturally tempted us to ask what are the fingerprints of the hierarchical structure on quantum interference effects.

This question was first investigated theoretically by Ketzmerick[92], who showed, using the semiclassical Landauer formula, that fluctuations of magneto-conductance, as a function of the weak magnetic field B, should resemble **fractional Brownian motion** and scale as $\langle (\Delta G)^2 \rangle \sim (\Delta B)^\gamma$, where γ is the power of the dwelling time probability which is assumed to be the same as the distribution of areas enclosed by the classical trajectories. The dimension of the fractal fluctuations, D_F, can then be written as $D_F = 2 - \frac{\gamma}{2}$. Since from classical dynamics we have $1 < \gamma < 2$, D_F should be between 1 and 1.5. This theoretical prediction has been confirmed qualitatively by experimental evidence, in gold nanowires, semiconductor microstructures, as well as in numerical simulations. However, it lacks quantitative confirmation. In addition, Micholich *et al.*'s experiments[90,91] discovered the exact self similarity of magneto-conductance in Sinai billiards with D_F larger than 1.5, which waits for a qualitatively more profound insight than Ketzmerick's. In the following we give some intriguing insight into this subject, by taking an alternative picture.

7.6.1 Harmonic Saddles as the Origin of Self-Similarity

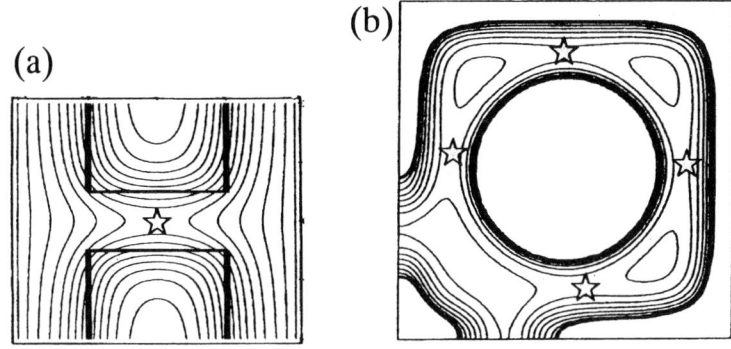

Fig. 7.9 Emergence of harmonic saddles (denoted by stars) in general soft-walled potentials. Potential contour for: (a) point contact; (b) single Sinai billiard system with leads.

While the above phenomenon has been attributed to the mixed phase-space structure in the underlying classical mechanics, we shall give here a very interesting periodic-orbit picture that the self-similar conductance fluctuations should be caused by the self-similar unstable periodic orbits which are generated through a sequence of isochronous pitchfork bifurcations of straight-line librating orbits oscillating towards harmonic saddles. The saddles are naturally created right at the point of contact with the leads or at certain places in the cavity as a consequence of the soft-wall confinement (see Fig. 7.9). Therefore this mechanism is able to explain all the self-similar magneto-conductance fluctuations in general soft-wall billiards. In contrast to Ketzmerick's theory which claims that a classical mixed phase space is necessary, we argue that, even in the fully chaotic phase space, the self-similar magneto-conductance fluctuations should be observed as long as the self-similar unstable periodic orbits are preserved.

Using the semiclassical Kubo formula (to be explained in the next chapter) for conductivity, Budiyono and Nakamura[93] gave a possible answer to the above question. We shall illustrate their idea below, by choosing a submicron-scale soft-walled multichannel billiard system whose linear dimension is much smaller than the typical electron mean free path and phase coherence length. In these single devices, the soft-wall boundary is experimentally essential. The exact self-similar magneto-conductance fluctuations will be shown due to the self-similar periodic orbits generated through a sequence of isochronous pitchfork bifurcations of straight-line librating orbits oscillating toward harmonic saddles.

The finite system under consideration is coupled to large baths of source and drain so that the energy levels inside the dots are no longer discrete, but have finite widths due to energy broadening. In fact, experimental results indicate that the phenomenon of fractal-like conductance fluctuations is observed when the average energy broadening is of the same order as the average energy level spacing. This fact together with the numerical evidence based on quantum calculations tempts us to apply the Kubo formula for conductivity rather than Landauer's formula to the present problem.

The oscillating part of the conductance is then approximated by that of the semi-classical Kubo formula in terms of periodic orbits as follows (see Chapter 8)

$$\delta G_{xx}(E, B) = \frac{2g_s e^2}{hV} \sum_{po} C_{xx}^{po} \frac{\Xi(T_{po}/\tau_\beta) F_{po}(\tau_{el})}{|\det(\tilde{M}_{po} - I)|^{1/2}}$$

$$\times \cos\left(\frac{S_{po}}{\hbar} - \frac{\mu_{po}}{2}\pi\right). \tag{7.48}$$

Here S_{po} is the action evaluated at the Fermi energy E_F; μ_{po} and $\tilde{M}_{po}$ are the Maslov index and stability matrix for each periodic orbit, respectively; g_s is a spin factor and V is the volume of the system. The temperature T selects only a few shortest periodic orbits that contribute to the trace through a damping factor

$$\Xi(T_{po}/\tau_\beta) = (T_{po}/\tau_\beta)/\sinh(T_{po}/\tau_\beta),$$

where T_{po} is the period of the periodic orbits and $\tau_\beta = \frac{\hbar}{\pi k T}$. Damping due to a finite mean free path is given by $F_{po}(\tau_{el}) = \exp(-T_{po}/2\tau_{el})$, where τ_{el} is the elastic scattering time. C_{xx}^{po} is the velocity–velocity correlation function of the periodic orbit, defined by

$$C_{xx}^{po} = \int_0^{T_{po}} dt \int_0^\infty dt' v_x(t) v_x(t + t') \exp(-t'/\tau_s). \tag{7.49}$$

On switching a small magnetic field B, we can assume that only the phase of the electron is changed, and the periodic orbits (the phase-space structure) remain unchanged. Then we can expand S_{po} up to the first order in B as follows:

$$S_{po}(E, B) = S_{po}(E, 0) + \frac{e}{c}\Theta_{po}B, \tag{7.50}$$

where Θ_{po} is the area enclosed by each periodic orbit. For a sufficiently weak magnetic field B, considering the contributions from $\pm\Theta_{po}$, i.e., from a pair of time-reversal symmetric orbits, one can rewrite the cosine terms as

$$2\cos\left(\frac{S_{po}(E, 0)}{\hbar} - \frac{\mu_{po}}{2}\pi\right)\cos\left(\frac{e}{\hbar c}\Theta_{po}B\right).$$

Now, let us suppose that, through some kind of bifurcations, we have a sequence of periodic orbits which are self-similar, and satisfy the following ideal relation:

$$\Theta_{po}^n = \lambda \Theta_{po}^{n-1}, \quad 0 < \lambda < 1. \tag{7.51}$$

Here Θ_{po}^n denotes the area enclosed by the periodic orbit generated at the n-th bifurcation. Then one can expect that fluctuations of the magneto-conductance should be

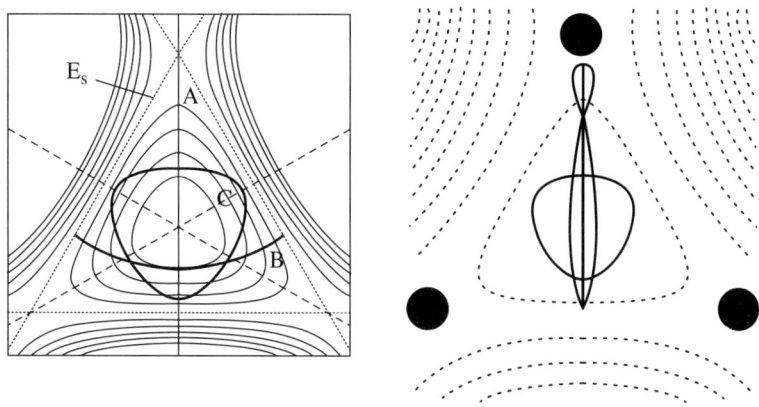

Fig. 7.10 Profile of a three-channel soft-wall billiard system modeled as a Henon–Heiles potential. Left: three shortest periodic orbits A, B and C (reproduced from[94]). Right: periodic orbits with nonvanishing area. Filled circles denote saddles. Mother orbit A which moves straight toward an upper saddle, the first child orbit R_5 with its shape like the number 8, which is generated from A through isochronous bifurcation, and irrelevant orbit C, which has nothing to do with bifurcations.

mimicked by the well-known Weierstrass function. The Weierstrass function is continuous but nowhere differentiable. It is commonly used to generate self-affine curves. In its simplest form, it can be written as follows:

$$f(x) = \sum_{n=0}^{\infty} a_n \cos(b_n \pi x),$$

where $0 < a \equiv a_n/a_{n-1} < 1$, and $b \equiv b_n/b_{n-1}$ is an odd integer that satisfies $ab > 1 + \frac{3}{2}\pi$. The frequency b_n must form an arithmatic progression, to guarantee the scaling in the x direction of the function, whereas the amplitude a_n should decrease quickly enough to make the function converge, and again the arithmetic decreasing progression of a_n guarantees the scaling in the y direction of the function. In the present problem, however, the summation will be truncated by the upper bound corresponding to Θ_{po}^0, i.e., the largest area enclosed by the periodic orbits, and this upper bound gives the smallest scale of the self-similar function.

The above relation is very common in open cavities with a soft-wall boundary. In fact any introduction of leads in a cavity will naturally create a harmonic saddle right on the point of contact, i.e., it gives a maximum to a longitudinal cut (parallel to each lead) and a minimum (which is harmonic) to a transversal cut. To make the

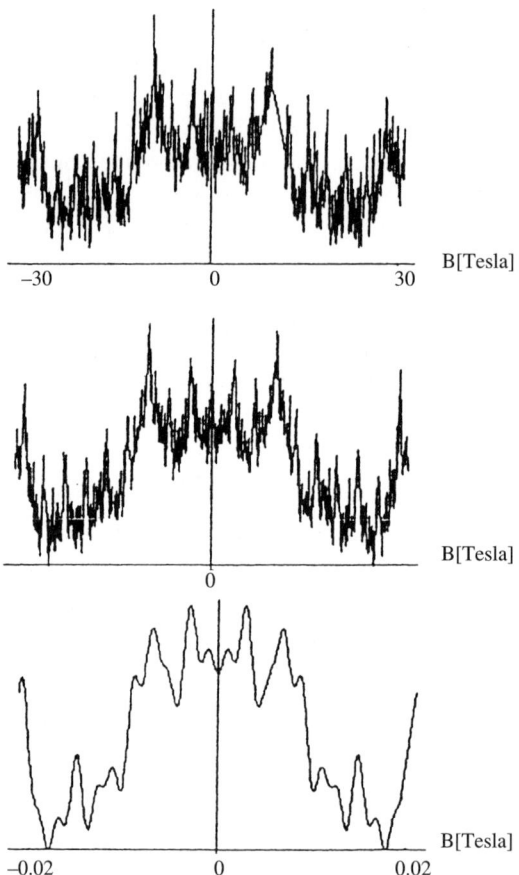

Fig. 7.11   Successive magnifications of fractal fluctuations of δG_{xx} around $B = 0$ tesla, for $T = 0.01$ K. The vertical coordinate is scaled in arbitrary units. The scale factor of each magnification in the horizontal direction is equal to 40 (reproduced from[93]).

picture clear, let us look at the well-known Henon–Heiles potential[2] shown in Fig. 7.10, which is given by the following equation:

$$V(x, y) = \frac{1}{2}(x^2 + y^2) + \epsilon(x^2 y - \frac{1}{3}y^3). \qquad (7.52)$$

The Henon–Heiles potential can be taken as a simple ideal model for a soft-wall triangular billiard system with three leads attached on its three edges. The harmonic saddles are obvious.

Fortunately, in this potential there are only three kinds of periodic orbits (see Fig. 7.10) with periods of the order of $\tau = 2\pi$ (i.e., the fundamental period of the harmonic-oscillator potential reached in the limit $\epsilon \to 0$), namely, a librating orbit A which will play the most important role, another librating orbit B which is unstable for all energy, and a rotating orbit C which is unstable for E_F sufficiently close to the saddle energy E_S (Brack[94]). The orbit B that has a vanishing area and the orbit C that shows no bifurcation will be swept away from our considerations. For E_F sufficiently close to E_S, the librating orbit A oscillates between stability and instability, undergoing an infinite number of isochronous bifurcations that accumulate at the saddle energy, where the period T_A becomes infinite. During the bifurcations two types of self-similar periodic orbits are generated. A sequence of librating orbits L_{2m} with vanishing area, which are born when the orbit A changes from unstable to stable, and a sequence of rotating orbits R_{2m-1} with a nonvanishing area, which are born when the orbit A changes from stable to unstable. Here, the lower index denotes the Maslov index of the periodic orbits. The important thing is that the bifurcation is isochronous, i.e., the period of the new-born periodic orbits is equal to the period of the mother orbit A at the bifurcation energy[94]. The periods of the children remain unchanged while the Fermi energy is increased and the period of a child is related to the period of its immediate brother (of the same type) as $T_{R_{2m+1}} = T_{R_{2m-1}} + \Delta T$, where $\Delta T = \dfrac{2\pi}{\omega_\perp}$ with $\omega_\perp^2 = \dfrac{\partial^2 V(x,y)}{\partial x^2}$ evaluated at the saddle point, measuring the **transversal curvature of the saddle**. The increasing energy is used to rescale the amplitude of the periodic orbit in the transverse direction. As the Fermi energy is increased, these children become more and more unstable, and the value of Tr $\tilde{M}_{po}$ converges to a certain value. For $E_F > E_S$, the mother orbit A disappears, but all its children still survive with the self-similarity preserved. A new librating periodic orbit which oscillates across the saddle appears, but is irrelevant because of its vanishing area.

7.6.2 Scaling Properties
We are only interested in the orbits with a nonvanishing area, R_{2n-1} (see Fig. 7.10). The closer they are born to E_S, the smaller their amplitude in the transversal direction. For sufficiently large E_F, the ratio of the amplitude of a child to that of its next brother is the same for all children and is equal to $\exp(-\pi/\omega_\perp)$. There is another scaling relation in the longitudinal direction, which relates the tip of the periodic orbit to the entire periodic orbit. However its contribution to the area is very small. Neglecting this contribution we obtain the scaling relation that relates the area of two immediate brothers as in (7.51) with $\lambda = \exp(-\pi/\omega_\perp)$.

In Fig. 7.11, using the periodic orbit C and the first eight children of the periodic orbit A, R_{2m-1}, $m = 3 \ldots 10$, we plot $\delta G_{xx}(B)$, evaluated at E_F a little larger than E_S. The exact self-similarity is obvious. The top of the figures are theoretically not allowed because the value of the magnetic field is too large to support the weak-field theory. Nevertheless, we plot them to show that the function obtained here is really an exact fractal function. However, there clearly exists a smallest scale at which the self-similarity is preserved finite. This corresponds to the largest area Θ_{po}^0, the area of the first child. These results indicate that the experimentally observed self-similar magneto-conductance fluctuations, which are up to now limited to only three orders of magnitude and have a lower cut at some large scale of magnetic field (10^{-3} tesla), are not only an observational matter but also have a physical basis. Since $E_F > E_S$, the system under consideration is in the fully chaotic state, and therefore the conclusion is that the fractal magneto-conductance fluctuations are not the fingerprints of the mixed phase space. Rather, they reflect the self-similar unstable periodic orbits of the system. Those periodic orbits that are born through period doubling bifurcations of the same straight-line librating orbits are ignored, for their periods increase quickly so that most of them are larger than the system phase-coherent time. Assuming the function obtained here to be approximated by the Weierstrass function, one can calculate the fractal dimension as $D_F = 2 - H$, where H is the Hurst exponent which is defined as the ratio of the logarithmic values of the scaling constants in the y and x directions. Here the dimension of the fractal magneto-conductance fluctuations is equal to 1.7, which is again violating the results predicted by Ketzmerick and supports the experimental results of Micholich *et al.*[90, 91].

As in transport in classical systems where the repeller plays a very important role, we argue as follows. Electrons in the cavity with initial conditions far enough from the cluster of periodic orbits will directly get out from the cavity and therefore do not contribute to the conductance fluctuations. In contrast to this, those electrons with initial conditions sufficiently close to one of the periodic orbits will be trapped for a while before getting out of the cavity. In doing so, the phase of the electrons acquires an additional term almost equal to $j\dfrac{e}{\hbar c}\Theta_{po}B$, where Θ_{po} is the area enclosed by the periodic orbits, and $j = 1, 2, 3 \ldots$ denotes its repetitions. Finally, the conductance fluctuations are determined by the interference of all those trajectories.

The ingredient of Budiyono and Nakamura's scenario[93] is the occurrence of the isochronous pitchfork bifurcations of the straight-line librating orbits oscillating towards a harmonic saddle. This is not specific to the model in Fig. 7.10. Actually, besides the above Henon–Heiles potential, pitchfork bifurcations also occur in other potentials of the same type. The periodic orbits born from straight-line librating orbits that oscillate towards a harmonic saddle or between two harmonic saddles (as in quartic Henon–Heiles potentials) are also self-similar with different values of the scaling constant λ. These various potentials correspond to various positions and widths of the leads or various shapes of the cavity. Budiyono and Nakamura's scenario should therefore remain valid in explaining all the fractal magneto-conductance fluctuation phenomena in general soft-wall billiards. Finally we expect that in a realistic system, due to geometrical constraints, there might be more than one λ in a single quantum billiard system.

The most interesting case is that of the exact self-similar magneto-conductance observed in Sinai billiards. In this case, the exact self-similarity provides a strong line of experimental evidence. We explain this qualitatively, leaving the quantitative evaluation for future work. Introducing a soft-wall antidot in the middle of a soft-wall square billiard system will create saddles at each side of the square where the inner circular wall and the outer wall have the closest distance. This is most clear in the result obtained by Taylor *et al.* The Poisson equation is calculated self-consistently by considering all the experimental constraints. Each corner of the billiard system can then be modelled as a kind of Henon–Heiles potential.

The success of application of the semiclassical Kubo formula to explain fractal magneto-conductance fluctuations in term of classical periodic orbits will drive one's mind to ask how this formula is closely related to the semiclassical Landauer approach based on nonclosed classical trajectories. Is it possible to derive a periodic-orbit description of the semiclassical conductance from the Landauer formula[95]? Interestingly, a similar question will be raised and resolved in the semiclassical approach for the conductance peaks in the Coulomb blockade phenomenon (see Chapter 9).

In this chapter, our explanation has been devoted to the ballistic weak localization, AAS oscillation in AB billiards and universal conductance fluctuations. These items are key words on the border between quantum chaos and ballistic quantum transport. Reflecting different dynamical features, chaotic or regular, quantum transport shows behaviors more fruitful than those in the diffusive regime. The global feature of classically chaotic open billiards is characterized by an exponential area distribution, dwelling time distribution, winding-number distribution and so on, which proved to affect the ballistic quantum interference effects seriously. Furthermore, the extended semiclassical theory that incorporates diffraction effects is used to resolve the problem involved in the early ballistic weak-localization theory and to achieve a quantitatively correct theory of ballistic conductance fluctuations. Finally we introduced the mysterious fractal structure implied in the experimentally-observed quantum oscillation in AB billiards. The "softwallness" of the billiards creates harmonic saddles right at the point of contact with the leads or at the middle of the sides of the square in Sinai billiards. The presence of the saddles is then responsible for the occurrence of the pitchfork bifurcations. The fractal-like fluctuations are due to the interference of the electron waves which move near the self-similar periodic orbits.

Several remarks should be made. First, one should not form a hasty conclusion that the scattering by billiard walls in the ballistic regime is equivalent to that by impurities in the diffusive regime. **The nature of chaos lies in neither (a) randomness nor (b) ergodicity. Rather it lies in (c) the orbital instability characterized with positive Lyapunov exponents and (d) the mixing property exemplified by stretching and folding transformations.** (c) and (d), which certainly lead to deterministic randomness and ergodicity, cannot be thoroughly replaced by (a) and (b). Second, for investigation of the rich features of ballistic quantum transport in mixed systems (of chaos and tori) and regular ones, neither the random matrix theory nor a super-matrix method (e.g., nonlinear sigma model) are effective, and instead the semiclassical theory and its extension become very helpful. In ballistic quantum transport where nonlinear dynamics of chaos and regular motions play a vital role, one is required to take off from the standard understanding of diffusive quantum transport.

8

LINEAR RESPONSE THEORY IN THE SEMICLASSICAL REGIME

In this chapter, a semiclassical theory of bulk quantum transport is explained. As understood in the previous chapter, submicron-scale quantum dots are phase-coherent, and their quantum transport properties are described on the basis of the Landauer formula. This formula exploits transmission and reflection components of scattering (S) matrices. By contrast, an antidot superlattice as a realization of Sinai billiards has a length scale that satisfies

period of superlattice ≪ mean free path and phase coherence length
≪ system size,

and the whole system is not always phase-coherent. In these circumstances, it is more suitable to have recourse to the Kubo formula that was originally constructed for bulk quantum transport. For illustration, we first consider a 2-d electron gas with magnetic field in the absence of antidots, namely, in an unpatterned conducting plane. We shall explain how the semiclassical theory nicely leads to the Shubnikov–de Haas oscillation in electrical conductivity. Then a general framework of semiclassical theory on antidot lattices is developed.

8.1 Realization of Sinai Billiards

The problem of **Sinai billiards** or **Lorentz gas** has received wide attention in the field of mathematical physics for the last few decades. This problem is to place a regular or irregular array of circular hard disks and pursue the complicated motion of a point particle moving between these disks. The central interest lies in giving a proof for the ergodicity and mixing which constitute the foundations of statistical mechanics. In 1991 and 1993, Weiss and coworkers[96, 97] drilled the interface of GaAs/AlGaAs semiconductor heterojunctions in which a high-mobility ($\mu \sim 1.0 \times 10^6 \text{cm}^2/\text{Vs}$) 2-d electron gas with carrier concentration $n_s \sim 2 \times 10^{11} \text{cm}^{-2}$ is confined, and, in analogy to Sinai billiards, they actually fabricated a square superlattice (with period $d = 200 \sim 300 \text{nm}$) of circular hard disks with radius $a \sim 50 \text{nm}$ (see Fig. 8.1). Applying a magnetic field perpendicular to the lattice plane, they measured quantum transport coefficients. This experiment converted a sophisticated billiard problem to a theme of real physics. The experiment of Weiss *et al.*, together with that of Marcus *et al.*[37] on single quantum dots (e.g., stadium billiards) (see Chapters 1 and 7), opened a new era of experimental research of quantum chaos.

Convex (or swollen) disks are called **quantum antidots** or simply **antidots**. A regular lattice of the disks is called an antidot superlattice, to distinguish it from conventional crystal lattices of atoms with lattice constant $\sim 0.1 \text{nm}$. In antidot lat-

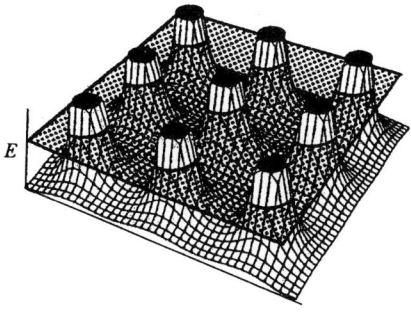

Fig. 8.1 Periodic potential in a 2-d antidot superlattice. The dotted surface represents the Fermi energy (reproduced from[98]).

tices, both the mean free path l_e and the phase-coherence length l_ϕ are of order $10\,\mu$m, much larger than the interdisk distance, and electrons show a ballistic motion between disks. Since the Fermi wavelength ($< 50\,$nm) is less than both the dot radius (a) and interdot distance (d), theoretical analyses based on the semiclassical approximation will be well justified. For an electron with Fermi energy $E_F \left(= \dfrac{\hbar^2 k_F^2}{2m_e} \right)$, its cyclotron radius is given by $R_c = \dfrac{\hbar k_F c}{eB}$, which can be made comparable to the interdot distance by tuning the magnetic field B.

The experiment of Weiss $et\ al.$ in 1991 measured the B-field dependence of electrical resistance against various aspect ratios $\left(\dfrac{a}{d} = \dfrac{1}{3} \sim \dfrac{1}{2} \right)$ at temperature $T \sim 1.5\,$K (see Fig. 8.2). The larger the aspect ratio, the higher the probability of an electron to collide with dots becomes. The inset on the upper right of Fig. 8.2 emphasizes the **commensurate orbits** of Weiss $et\ al.$ that are commensurate with the periodicity of a superlattice, encircling one or several antidots. The commensurate orbits more or less suppress electronic transmission and enhance the resistance. Resistance peaks can be explained in terms of these orbits by using the Drude conductivity formula. In fact, in the small aspect-ratio case, the experiment captured the peaks associated with commensurate orbits encircling up to 21 antidots (see curve I). With increase of the aspect ratio, however, such a clear interpretation breaks down except for the peak arising from the fundamental orbit with radius $2R_c = d$.

To see in detail the quantum-mechanical effects on the spectrum in the case of $\dfrac{a}{d} = \dfrac{1}{2}$ (curve III in Fig. 8.2), Weiss $et\ al.$ measured the resistance ρ_{xx} at a much lower temperature $T \sim 0.4\,$K in 1993. Then a new quantum oscillation, which is absent at the temperature $T \sim 1.5$K, shows up in the lower-field side ($2R_c > d$) of the fundamental commensurability peak (see Fig. 8.3). This oscillation has period $\Delta B \sim 0.105\,$T $\sim \dfrac{h}{ed^2}$, and can be attributed neither to the Shubnikov–de Haas

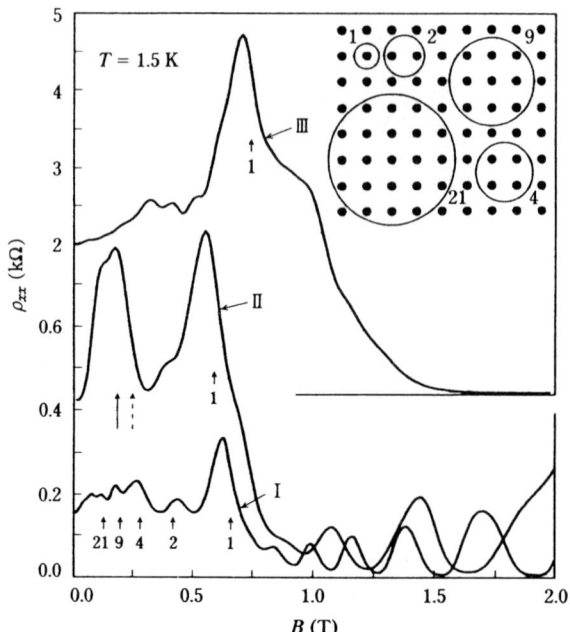

Fig. 8.2   Magneto-resistance ρ_{xx} at $T = 1.5\,\mathrm{K}$. The aspect ratio a/d increases from the bottom curve through the top (in order of I, II, III). Labels $1, 2, 9, 4, 21$ in the upper right inset denote the number of antidots enclosed by incommensurate orbits. Reprinted from[98].

oscillation (periodic in B^{-1}) due to the Landau-level structure in the high-field regime nor to the Aharonov–Bohm oscillation in the low-field regime where a perturbational treatment should work well. The successful interpretation will need a knowledge of the semiclassical theory of bulk quantum transport, which will be described hereafter.

8.2   Semiclassical Shubnikov–de Haas Oscillation

Antidot superlattices have three length scales: d (period of superlattice), l_e (mean free path) and l_ϕ (phase coherence length), which typically satisfy

$$d \ll l_e, l_\phi \ll \text{size of whole superlattice}.$$

The inequality means that antidot lattices as a whole are not always phase-coherent and that the Kubo formula, which is originally used for the description of bulk transport, is convenient to describe their quantum transport. In this chapter, we start with a 2-d electron gas with magnetic field in the absence of antidot patterns, and derive

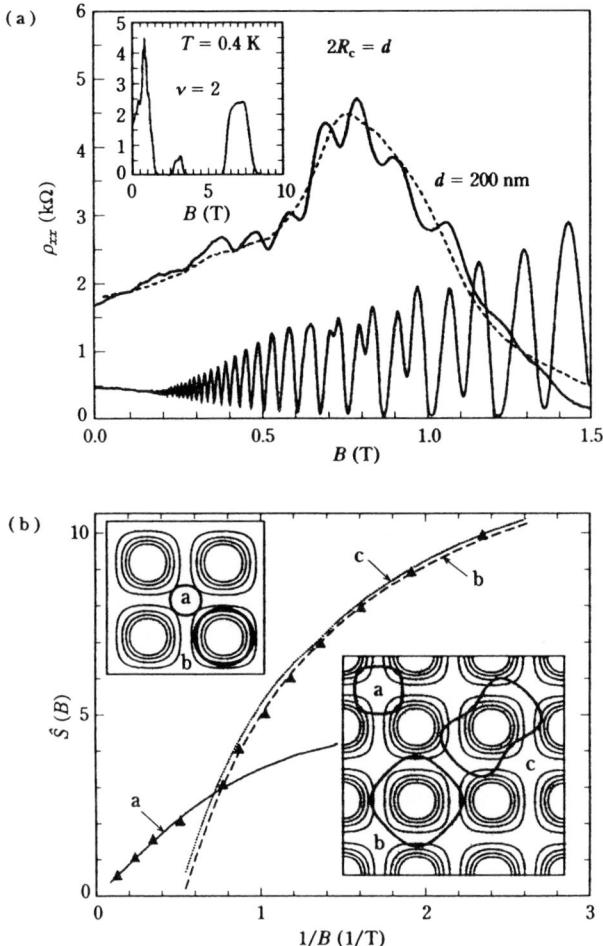

Fig. 8.3 (a) Measured ρ_{xx}. Upper curves (solid and broken lines correspond to $T = 0.4\,\text{K}$ and $4.7\,\text{K}$, respectively): low-field oscillations in the presence of antidot pattern. A more global spectrum in a field up to $10\,\text{T}$ is given in the inset. Lower curve: Shubnikov–de Haas oscillation in the absence of antidot pattern. (b) Values of effective action $\widehat{S}(B)$. Solid, broken and dotted lines correspond to orbits a, b, c in the inset, respectively. Dark triangles are values of $1/B$ at which ρ_{xx} shows minima. The inset on the upper left (lower right) indicates orbits in the presence of a B field with $1/B = 0.6\ \text{T}^{-1}$ ($1/B = 2.7\ \text{T}^{-1}$). Reprinted from[98].

the Shubnikov–de Haas oscillation in electrical conductivity within the framework of semiclassical theory together with knowledge of nonlinear dynamics. This will be an instructive chart for constructing the semiclassical transport theory of antidot lattices in the following sections. The scheme in this section is based on the work by Hackenbroich and Oppen[59].

The Kubo formula for the longitudinal conductivity is

$$\sigma_{xx} = \frac{e^2 \pi \hbar}{V} \, \mathrm{Tr} \left\{ \hat{v}_x \delta_\Gamma (E_\mathrm{F} - H) \hat{v}_x \delta_\Gamma (E_\mathrm{F} - H) \right\}, \tag{8.1}$$

where V is the area (or volume in the 3-d case) of the whole system, and both $\hat{v}_x$ and H are operators: $\hat{v}_x$ stands for the x component of the velocity, and

$$H = \frac{1}{2m_e} \left(\boldsymbol{p} + \frac{e}{c} \boldsymbol{A}(\boldsymbol{r}) \right)^2$$

is the Hamiltonian for the 2-d electron gas with effective electron mass m_e in the magnetic field perpendicular to the system's plane. (For the above Hamiltonian form, see the footnote in Section 6.1.) The vector potential $\boldsymbol{A}$ is given by $\boldsymbol{A} = \left(-\frac{1}{2} By, \frac{1}{2} Bx \right)$ in a symmetric gauge. (8.1) can be rewritten in terms of Green functions. First one should note that the δ function can be expressed with the use of retarded ($+$) and advanced ($-$) Green functions $G^{\pm}(E) = [E - H \pm i \Gamma]^{-1}$ as follows:

$$\delta_\Gamma (E_\mathrm{F} - H) = -\frac{1}{2\pi i} [G^+(E_\mathrm{F}) - G^-(E_\mathrm{F})]. \tag{8.2}$$

In this treatment, the subscript Γ has a nonzero value $\Gamma = \dfrac{\hbar}{2\tau_{el}}$ with elastic scattering time τ_{el}, indicating a level broadening effect due to the system's small impurities.

Substituting (8.2) together with the coordinate representation for Green functions

$$G^{\pm}_{\boldsymbol{r}, \boldsymbol{r}'}(E) = \langle \boldsymbol{r} | G^{\pm}(E) | \boldsymbol{r}' \rangle$$

into (8.1), one finds

$$\sigma_{xx} = \sigma_{xx}^{-+} + \sigma_{xx}^{++} + \sigma_{xx}^{--}, \tag{8.3}$$

where σ_{xx}^{-+} has the form

$$\sigma_{xx}^{-+} = 2 \frac{e^2 \pi \hbar}{(2\pi)^2 V} \int d^2 r \, d^2 r' \left[\hat{v}_x G^-_{\boldsymbol{r}, \boldsymbol{r}'}(E_\mathrm{F}) \right] \left[\hat{v}'_x G^+_{\boldsymbol{r}', \boldsymbol{r}}(E_\mathrm{F}) \right] \tag{8.4}$$

with $\hat{v}_x$ and $\hat{v}'_x$ defined by

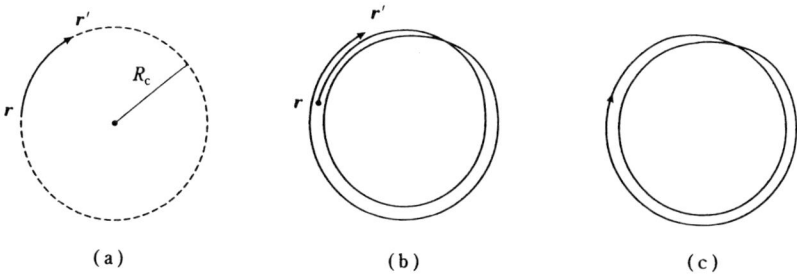

Fig. 8.4 Incomplete circular orbit (radius R_c): (a) irreducible part; (b) irreducible part + double repetitions of complete Larmor orbit; (c) double repetitions of complete Larmor orbit (primitive periodic orbit).

$$m_e \hat{v}_x \equiv \frac{\hbar}{i} \nabla_x + \frac{e}{c} A_x(\boldsymbol{r}), \qquad m_e \hat{v}'_x \equiv \frac{\hbar}{i} \nabla_{x'} + \frac{e}{c} A_x(\boldsymbol{r}').$$

$\sigma^{\pm\pm}_{xx}$ are given in a similar way using the same-type Green functions ($G^\pm G^\pm$). In the semiclassical limit, however, $\sigma^{\pm\pm}_{xx}$ are actually suppressed because of a violent oscillation of integrands due to nonvanishing phases, and σ^{-+}_{xx} alone can survive in (8.3).

As described in Chapter 5, the semiclassical limit of $G^+_{\boldsymbol{r}',\boldsymbol{r}}(E_F)$ is a sum of contributions from classical orbits (with energy E_F) from $\boldsymbol{r}$ to $\boldsymbol{r}'$. For an electron in a magnetic field, the corresponding orbits are **incomplete circular orbits** from $\boldsymbol{r}$ to $\boldsymbol{r}'$. More precisely, the incomplete Larmor orbits, with radius $R_c \left(= \dfrac{\hbar k_F c}{eB} \right)$ and chord length equal to $|\boldsymbol{r} - \boldsymbol{r}'|$, give a nonvanishing contribution to the retarded Green function $G^+_{\boldsymbol{r}',\boldsymbol{r}}(E_F)$ (see Fig. 8.4(a)), which reads (Feynman and Hibbs[3])

$$G^+_{\boldsymbol{r}',\boldsymbol{r}}(E) = \frac{2\pi}{(2\pi i \hbar)^{\frac{3}{2}}} \sum_{n=0}^{\infty} \sum_{\alpha=\pm} A_{n,\alpha} \exp\left(\frac{i}{\hbar} S_{n,\alpha} - \frac{i\pi}{2} \mu_{n,\alpha} \right), \qquad (8.5)$$

where

$$A_{n,\alpha} = \frac{m_e}{2} \left(\frac{2\omega_c}{E |\sin(\omega_c T_{n,\alpha})|} \right)^{1/2} \exp\left(-\frac{T_{n,\alpha}}{2\tau_{el}} \right) \qquad (8.6)$$

with $\omega_c = \dfrac{eB}{m_e c}$ the cyclotron frequency. The amplitude factor (8.6), which reflects the orbital instability (see also (8.17) in the next section), is now modified by the Born factor due to the aforementioned weak disorder (τ_{el} is the elastic scattering time).

Each incomplete orbit from $\boldsymbol{r}$ to $\boldsymbol{r}'$ is decomposed into two parts: the irreducible one with its *arc length* (l_{ir}) less than the circumference of the complete Larmor orbit ($2\pi R_c$) and the periodic one, namely, integer (n) multiples of the complete Larmor orbit (see Fig. 8.4(b)). Further, for a given pair $\boldsymbol{r}$ and $\boldsymbol{r}'$, there exist two kind of orbits, one with the irreducible part shorter than half the circumference ($0 < l_{ir} \leq \pi R_c$) and the other with $\pi R_c < l_{ir} < 2\pi R_c$. The former and latter orbits are denoted as $\alpha = -$ and $\alpha = +$, respectively. The flight time of each incomplete orbit is now

$$T_{n,\alpha} = t_\alpha + \frac{2\pi n}{\omega_c}, \tag{8.7}$$

with t_α being the time for the irreducible part:

$$t_- = \frac{2}{\omega_c} \sin^{-1} \left(\frac{|\boldsymbol{r} - \boldsymbol{r}'|}{2R_c} \right), \qquad t_+ = \frac{2\pi}{\omega_c} - t_-. \tag{8.8}$$

The concrete form of the action $S_{n,q}$ in (8.5) is

$$S_{n,\alpha}(E) = \int_{\boldsymbol{r}}^{\boldsymbol{r}'} \boldsymbol{p} \cdot d\boldsymbol{r}$$

$$= ET_{n,\alpha} + \frac{m_e \omega_c}{2} \left[\frac{|\boldsymbol{r} - \boldsymbol{r}'|^2}{2} \cot \left(\frac{\omega_c T_{n,\alpha}}{2} \right) - (xy' - x'y) \right]. \tag{8.9}$$

Both the terms on the last line in (8.9) precisely include the effect of orbital bending. The Maslov index $\mu_{n,\alpha}$ is the number of sign changes in one component of momentum $\boldsymbol{p}$, say p_x: $\mu_{n,-} = 2n$ and $\mu_{n,+} = 2n + 1$.

On the other hand, the advanced Green function is the complex conjugation of (8.5), i.e.,

$$G^-_{\boldsymbol{r},\boldsymbol{r}'}(E) = \left[G^+_{\boldsymbol{r}',\boldsymbol{r}}(E) \right]^*,$$

indicating that both $G^+_{\boldsymbol{r}',\boldsymbol{r}}(E)$ and $G^-_{\boldsymbol{r},\boldsymbol{r}'}(E)$ are the sum of the identical contributions due to paths from $\boldsymbol{r}$ to $\boldsymbol{r}'$ except for the sign difference in the phase of each contribution.

With use of (8.5) in (8.3) and (8.4), and taking the semiclassical limit that keeps only the terms diagonal ($\alpha_1 = \alpha_2 = \alpha$) in the irreducible part, one obtains

$$\sigma_{xx} = \frac{1}{V} \left(\frac{e}{2\pi\hbar} \right)^2 \sum_{n_1=0}^{\infty} \sum_{n_2=0}^{\infty} \sum_{\alpha=\pm} \int d^2\boldsymbol{r} \int d^2\boldsymbol{r}' (v_\alpha)_x (v'_\alpha)_x$$

$$\times A^*_{n_1,\alpha} A_{n_2,\alpha} \exp \left\{ \frac{i}{\hbar} (S_{n_2,\alpha} - S_{n_1,\alpha}) - \frac{i\pi}{2} (\mu_{n_2,\alpha} - \mu_{n_1,\alpha}) \right\}, \tag{8.10}$$

where the operator $\hat{v}_x$ has now become c-number after the following rule,

$$\hat{v}_x \exp \left(\frac{i}{\hbar} S_{n,\alpha} \right) = \frac{1}{m_e} \left(\frac{\hbar}{i} \nabla_x + \frac{e}{c} A_x(\boldsymbol{r}) \right) \exp \left(\frac{i}{\hbar} S_{n,\alpha} \right)$$

$$= -(v_\alpha)_x \exp \left(\frac{i}{\hbar} S_{n,\alpha} \right),$$

and similarly $\hat{v}'_x$ is reduced to $(v'_\alpha)_x$. Since the difference between the two kinds of incomplete orbits in Figs. 8.4(a) and 8.4(b) is nothing but repetitions of the complete Larmor orbit, the difference between the actions, $S_{n_2,\alpha} - S_{n_1,\alpha}$, is equal to an integer multiple of the action for the primitive periodic orbit (complete Larmor orbit) with no dependence on initial (r) and final (r') points. The difference between the Maslov indices also obeys the same rule.

The integrand in (8.10) depends only on a single variable $|r - r'|$. Noting further that $|r - r'|$ is determined by t_α through (8.7) and (8.8), the integration over r, r' is reduced to that over t_α. Including the whole incomplete orbits with $\alpha = \pm$, this reduction is made as follows:

$$
\int_V d^2r \int_{|r - r'| \leqq 2R_c} d^2r'
$$

$$
= 2\pi V \int_0^{2R_c} |r - r'| \, d|r - r'|
$$

$$
= 2\pi V \omega_c R_c^2 \left(\int_0^{\pi/\omega_c} dt_- \, |\sin(\omega_c t_-)| + \int_{\pi/\omega_c}^{2\pi/\omega_c} dt_+ \, |\sin(\omega_c t_+)| \right). \tag{8.11}
$$

At the same time, we find

$$
(v_q)_x (v'_q)_x = v_x(0) v_x(t).
$$

After the above tricks, σ_{xx} in (8.10) becomes

$$
\sigma_{xx} = \frac{e^2 m_e}{2\pi \hbar^2} \int_0^{2\pi/\omega_c} dt v_x(0) v_x(t) \exp\left(-\frac{t}{\tau_{el}} \right)
$$

$$
\times \sum_{n_1=0}^{\infty} \sum_{n_2=0}^{\infty} \exp\left\{ -\frac{\pi}{\omega_c \tau_{el}} (n_1 + n_2) \right\}
$$

$$
\times \cos\left\{ 2\pi \left(\frac{E_F}{\hbar \omega_c} - \frac{1}{2} \right) (n_2 - n_1) \right\}. \tag{8.12}
$$

Further, defining the difference $p = |n_2 - n_1|$ and taking the summation over the sum $n_2 + n_1$ in advance, one reaches the final step,

$$
\sigma_{xx} = \frac{2e^2}{h} \left(\frac{E_F}{2\pi \hbar \omega_c} \right) C(v_x, v_x)
$$

$$
\times \left[1 + 2 \sum_{p=1}^{\infty} \exp\left(-\frac{p\pi}{\omega_c \tau_{el}} \right) \cos\left\{ 2\pi p \left(\frac{E_F}{\hbar \omega_c} - \frac{1}{2} \right) \right\} \right]
$$

$$
\equiv \sigma_{xx}^{cl} + \delta \sigma_{xx}. \tag{8.13}
$$

Here

$$
C(v_x, v_x) = \frac{1}{R_c^2} \int_0^{2\pi/\omega_c} dt \int_0^{\infty} dt' \, v_x(t) v_x(t + t') \exp\left(-\frac{t'}{\tau_{el}} \right)
$$

$$(8.14)$$

is the velocity correlation function, whose integrated value leads to the smooth part of σ_{xx} in (8.13), namely

$$\sigma_{xx}^{\text{cl}} = \frac{e^2 \tau_{el} n_e}{m_e} \left[1 + (\omega_c \tau_{el})^2\right]^{-1},$$
$$(8.15)$$

where n_e is the electron density. In 2-d systems, $n_e = g(E_{\text{F}})E_{\text{F}}$ with constant density of states $g(E_{\text{F}}) = \dfrac{m_e}{2\pi\hbar^2}$.

The spirit of semiclassical theory lies in describing quantum transport by using knowledge of nonlinear classical dynamics. The formula (8.13) is the result of semiclassical theory applied to electrical conductivity for a 2-d electron gas in a magnetic field. The smooth ($p = 0$) part σ_{xx}^{cl} stands for the classical Drude conductivity extended to systems with finite magnetic field; the oscillating ($p \neq 0$) part $\delta\sigma_{xx}$, which has appeared thanks to the constructive interference between a pair of open orbits, reproduces the Shubnikov–de Haas oscillation that is periodic in B^{-1}. The important point is the emergence of a primitive periodic orbit (complete Larmor orbit) and its repetitions as a result of **quantum interference**.

8.3   Semiclassical Kubo Formula in Antidot Superlattices

It is straightforward to extend the above theory to patterned systems. Let us move on to the semiclassical theory of quantum transport in chaotic antidot superlattices. Many periodic orbits will be relevant here, rather than only a single one as in the previous section. For an antidot square lattice with periodic potential $U(\boldsymbol{r})$ (period d), the Hamiltonian is

$$H = \frac{1}{2m_e}\left(\boldsymbol{p} + \frac{e}{c}\boldsymbol{A}\right)^2 + U(\boldsymbol{r}).$$

The semiclassical path-integral representation of the retarded Green function is given as a sum of contributions associated with paths γ from $\boldsymbol{r}$ to $\boldsymbol{r}'$:

$$G^+_{\boldsymbol{r}',\boldsymbol{r}}(E) = \frac{2\pi}{(2\pi i\hbar)^{3/2}} \sum_{\gamma} A_\gamma(\boldsymbol{r}',\boldsymbol{r})$$
$$\times \exp\left\{\frac{i}{\hbar}S_\gamma(\boldsymbol{r}',\boldsymbol{r};E) - i\frac{\pi}{2}\mu_\gamma\right\}.$$
$$(8.16)$$

In (8.16), the action is defined as

$$S_\gamma(\boldsymbol{r}',\boldsymbol{r};E) = \int_{\boldsymbol{r}(\gamma)}^{\boldsymbol{r}'} \boldsymbol{p} \cdot d\boldsymbol{r},$$

and the amplitude due to quantum fluctuations around each classical path is

$$A_\gamma(\boldsymbol{r}', \boldsymbol{r}) = \exp\left(-\frac{T_\gamma}{2\tau_{el}}\right) \begin{vmatrix} \dfrac{\partial^2 S_\gamma}{\partial r_i' \partial r_j} & \dfrac{\partial^2 S_\gamma}{\partial E \partial r_j} \\ \dfrac{\partial^2 S_\gamma}{\partial r_i' \partial E} & \dfrac{\partial^2 S_\gamma}{\partial^2 E} \end{vmatrix}^{1/2} \tag{8.17}$$

where $i, j = x, y$ and the multiplicative Born factor is due to weak disorder. The semiclassical representation for the advanced Green function $G^-_{\boldsymbol{r}, \boldsymbol{r}'}(E)$, which is a complex conjugation of $G^+_{\boldsymbol{r}', \boldsymbol{r}}(E)$, consists of contributions due to the same paths from $\boldsymbol{r}$ to $\boldsymbol{r}'$ as for $G^+_{\boldsymbol{r}', \boldsymbol{r}}(E)$ except for the reversed sign in each phase. (Recall a similar statement below (8.9).) A generalization of the previous section to antidot systems and the derivation of the semiclassical Kubo formula were made by Hackenbroich and Oppen[59] and independently by Richter, Ullmo and Jalabert[60]. We shall follow the former.

8.3.1 Drude Conductivity

Using (8.16) in (8.3) and (8.4) and picking up the phase-cancelling term, we have the the smooth part

$$\sigma^{cl}_{xx} \cong \frac{1}{V}\left(\frac{e}{2\pi\hbar}\right)^2 \int d^2r d^2r' \sum_\gamma v_x v_x' A_\gamma^2(\boldsymbol{r}', \boldsymbol{r}), \tag{8.18}$$

where we have again resorted to the equality below (8.4). (8.18) can be reduced to a more familiar expression as follows. First, let us insert the identity

$$1 = \int dE \delta(E - E_F)$$

in the integral on the right-hand side of (8.18) and transform the integration variables from $\boldsymbol{r}, \boldsymbol{r}', E$ to $\boldsymbol{r}, \boldsymbol{p}$ (initial momentum), t (time along a path) with their relationship

$$d^2r d^2p dt = d^2r d^2r' dE \sum_\gamma \begin{vmatrix} \dfrac{\partial p_i}{\partial r_j'} & \dfrac{\partial p_i}{\partial E} \\ \dfrac{\partial t}{\partial r_j'} & \dfrac{\partial t}{\partial E} \end{vmatrix}. \tag{8.19}$$

Noting that $p_i = -\dfrac{\partial S}{\partial r_i}$, $t = \dfrac{\partial S}{\partial E}$, the Jacobian in (8.19) turns out to be equal to $A_\gamma^2(\boldsymbol{r}', \boldsymbol{r})$ except for the Born factor (see (8.17)). Second, with the use of the density of states $g(E_F)$, let us define the phase-space average of an arbitrary observable $Q(\boldsymbol{p}, \boldsymbol{r})$ as

$$\langle Q(\boldsymbol{p}, \boldsymbol{r})\rangle_{\boldsymbol{p}, \boldsymbol{r}} = \frac{1}{h^2 V g(E_F)} \int d^2r d^2p Q(\boldsymbol{p}, \boldsymbol{r}) \delta(E_F - H(\boldsymbol{p}, \boldsymbol{r})). \tag{8.20}$$

In this way (8.18) is rewritten as

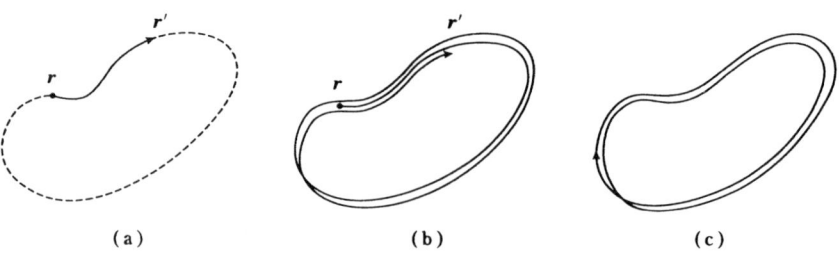

Fig. 8.5   (a), (b) Open orbits γ_1, γ_2 from r to r' satisfying (8.23). (c) An integer multiple of the primitive periodic orbit $\gamma(= \gamma_2 - \gamma_1)$ leading to (8.24).

$$\sigma_{xx}^{cl} \cong e^2 g(E_F) \int_0^\infty dt \, \langle v_x(0) v_x(t) \rangle_{p,r} \exp\left(-\frac{t}{\tau_{el}}\right), \tag{8.21}$$

and one comes to realize that, in antidot lattices, the smooth part of the semiclassical conductivity gives rise to the familiar Drude formula.

8.3.2   Quantum Correction

In contrast, the interference among a pair of open paths gives rise to a quantum correction

$$\delta\sigma_{xx} = \frac{e^2}{h^2 V} \int d^2r d^2r' \sum_{\gamma_1 \neq \gamma_2} (v_1)_x (v'_2)_x A_{\gamma_1}(r', r) A_{\gamma_2}(r', r)$$
$$\times \exp\left\{ \frac{i}{\hbar}\left(S_{\gamma_2}(r', r; E_F) - S_{\gamma_1}(r', r; E_F) \right) - \frac{i\pi}{2}\left(\mu_{\gamma_2} - \mu_{\gamma_1} \right) \right\}. \tag{8.22}$$

In the semiclassical limit, the above double intergrals can be evaluated by a saddle-point method: the saddle-point condition is

$$\nabla_r (S_{\gamma_2} - S_{\gamma_1}) \equiv p_1 - p_2 = 0,$$
$$\nabla_{r'} (S_{\gamma_2} - S_{\gamma_1}) \equiv p'_2 - p'_1 = 0, \tag{8.23}$$

indicating open orbits γ_1 and γ_2 to have equal momentum at each of the initial (r) and final (r') points. This condition is satisfied in two cases: (1) γ_1 and γ_2 are the same orbits; (2) both γ_1 and γ_2 lie on a common primitive periodic orbit γ (see Fig. 8.5). The case (1) cancels the phase in (8.22) and already leads to the Drude conductivity, so we shall not consider it further. On the other hand, the case (2) is essential here, giving phase-interference terms. The nonvanishing phase is given by the difference between actions, which is an integer (p) multiple of S_γ for the primitive periodic orbit γ as

$$S_{\gamma_2}(\mathbf{r}', \mathbf{r}; E_{\mathrm{F}}) - S_{\gamma_1}(\mathbf{r}', \mathbf{r}; E_{\mathrm{F}}) = p S_{\gamma}(E_{\mathrm{F}}). \tag{8.24}$$

Fluctuations around the saddle-point solution (8.23) and (8.24) are evaluated by changing the integration variables $\mathbf{r}, \mathbf{r}'$ in (8.22) as follows: first, $\mathbf{r}$ is represented as $\mathbf{r} = (z, y)$ with z and y parallel and perpendicular to γ, respectively, and similarly for $\mathbf{r}'$ with $0 < z < L_\gamma, 0 < z' - z < L_\gamma$. Then, moving to time variables as $z = vt, z' = vt'$, the integration over y, y', t, t' yields

$$\delta\sigma_{xx} = \frac{2e^2}{hd^2} \sum_\gamma T_\gamma C_\gamma(v_x, v_x) \sum_{p=1}^{\infty} \exp\left(-\frac{pT_\gamma}{2\tau_{el}}\right) \frac{\cos\left(\frac{pS_\gamma(E_{\mathrm{F}})}{\hbar} - p\mu_\gamma\frac{\pi}{2}\right)}{|\det(\tilde{M}_\gamma^p - I)|^{1/2}}. \tag{8.25}$$

Several notions should be addressed to (8.25). Due to the translational symmetry of the antidot (square) lattice (with lattice constant d), each periodic orbit appears $\dfrac{V}{d^2}$ times, so the periodic orbit sum in (8.25) is given in the form

$$\left(\frac{V}{d^2}\right) \times (\text{sum over different kinds of periodic orbits}).$$

The stability matrix $\tilde{M}_\gamma$ characterizes how the initial deviation $\delta\mathbf{x} = \begin{pmatrix} \delta y \\ \delta p_y \end{pmatrix}$ perpendicular to γ is magnified after a one-period excursion, and is defined as

$$\delta\mathbf{x}' = \tilde{M}_\gamma \delta\mathbf{x}.$$

The remaining factor in (8.25) is time-integration of the velocity correlation function,

$$C_\gamma(v_x, v_x) = \frac{1}{T_\gamma} \int_0^{T_\gamma} dt \int_0^\infty dt' v_x(t) v_x(t + t') \exp\left(-\frac{t'}{\tau_{el}}\right). \tag{8.26}$$

The quantum correction (8.25) is a generalization of the semiclassical expression for Shubnikov–de Haas oscillation in (8.13) to quantum transport in antidot lattices, and is a sum over an infinite number of terms due to periodic orbits, each having the amplitude factor determined by the stability matrix. (8.25) formally accords with the Gutzwiller trace formula (5.29) for the oscillating part of the density of states, aside from two factors (i.e., the damping factor and velocity fluctuations). Interestingly, in contrast with Gutzwiller's formula, periodic orbits in (8.25) come from the quantum interference between a pair of open orbits.

In the high-field regime where the Larmor radius $(R_c) \ll$ lattice constant (d), (8.25) reduces to (8.13), leading to the the Shubnikov–de Haas oscillation. In the intermediate-field regime that $R_c \sim d$, bent orbits demonstrate orbit bifurcations one after another, so $\delta\sigma_{xx}(B)$ shows a complicated B dependence. Finally, in the vicinity of zero field, one can derive the weak localization arising from time-reversal symmetry (see Section 7.3).

8.4   Effect of Finite Temperature and Spin

It is not difficult to proceed to the **semiclassical theory at finite temperatures** by generalizing the theory at zero temperature in the previous section. At finite temperatures, the effect of energy broadening at the Fermi energy modifies (8.1) as follows:

$$\sigma_{xx}(T, B, \mu)$$
$$= -\frac{e^2\pi\hbar}{V} \int_0^\infty dE \frac{\partial f(E-\mu)}{\partial E} \text{Tr}\left\{\hat{v}_x \delta_\Gamma(E-H)\hat{v}_x \delta_\Gamma(E-H)\right\}. \tag{8.27}$$

Here $f(E-\mu)$ is the Fermi distribution

$$f(E-\mu) = \left\{1 + \exp\left(\beta(E-\mu)\right)\right\}^{-1},$$

with chemical potential μ and inverse temperature $\beta \left(= \dfrac{1}{k_B T}\right)$. The zero-temperature result (8.25) points out that the energy dependence of the quantum correction is involved in the amplitude and phase of each periodic-orbit contribution. To obtain a finite temperature version of each contribution, an integration of the type

$$Y(T) = -\int_0^\infty dE \frac{\partial f(E-\mu)}{\partial E} A_\gamma(E) \cos\left(\frac{p}{\hbar}S_\gamma(E) - \frac{\pi}{2}p\mu_\gamma\right) \tag{8.28}$$

should be carried out. This can be done in a complex integral, noting that $f(E-\mu)$ in the complex energy plane has first-order poles at

$$E_n = \mu + \frac{(2n+1)\pi i}{\beta}$$

with $n = 0, \pm1, \pm2, \cdots$. The procedure of the integration is parallel to that from (5.41) through (5.49) in Chapter 5 for the oscillating part of the density of states, and the result is

$$Y(T) = \Xi\left(\frac{pT_\gamma}{t_T}\right) A_\gamma(\mu) \cos\left(\frac{p}{\hbar}S_\gamma(\mu) - \frac{\pi}{2}p\mu_\gamma\right), \tag{8.29}$$

where $\Xi\left(\dfrac{pT_\gamma}{t_T}\right)$ defined by

$$\Xi\left(\frac{pT_\gamma}{t_T}\right) = \frac{\dfrac{pT_\gamma}{t_T}}{\sinh\left(\dfrac{\pi pT_\gamma}{t_T}\right)} \tag{8.30}$$

is a damping factor to suppress the contribution of periodic orbits with period longer than the cutoff time $t_T = \dfrac{\hbar\beta}{\pi}$. Therefore the quantum correction to the conductivity at finite temperatures is given by

$$\delta\sigma_{xx}(T, B, \mu) = \frac{2e^2}{hd^2} \sum_\gamma T_\gamma C_\gamma(v_x, v_x) \sum_{p=1}^\infty \exp\left(-\frac{pT_\gamma}{2\tau_{el}}\right)$$

$$\times \Xi \left(\frac{pT_\gamma}{t_T} \right) \frac{\cos \left(\frac{pS_\gamma(E_{\mathrm{F}})}{\hbar} - p\mu_\gamma \frac{\pi}{2} \right)}{|\det(\tilde{M}_\gamma^p - I)|^{1/2}}. \qquad (8.31)$$

When the spin degrees of freedom are taken into consideration, one should note that the Fermi energy differs between up- and down-spins, i.e., $E_{\mathrm{F}} \to E_{\mathrm{F}} \pm \mu_{\mathrm{B}} B$ with Bohr magneton $\mu_{\mathrm{B}} = \dfrac{e\hbar}{2m_e c}$. Combining $\delta\sigma_{xx}(T, B, \mu)$ for up- and down-spins, and expanding the actions up to the term linear in B, the improved quantum correction is straightforward: it is similar to (8.31), but each term in this formula is multiplied by the factor $g_s(B) = 2\cos\left(\dfrac{pT_\gamma \mu_{\mathrm{B}} B}{\hbar} \right)$.

8.5 Interpretation of Experiments

In experimentally fabricated antidot lattices with soft-wall antidots (see Fig. 8.1), classical electron dynamics exhibits the coexistence of chaos and KAM tori in phase space. This superlattice, although not exactly corresponding to Sinai billiards, becomes fully chaotic in the presence of a weak magnetic field with its cyclotron radius satisfying $R_c \geq 2d$. In these circumstances, the isolated periodic orbits contribute to (8.31), and at finite temperatures actually short periodic orbits alone should be taken into consideration.

Weiss et $al.$ attempted to explain the experimental result by applying (8.31) to soft-wall antidots. The maxima in the conductivity $\sigma_{xx}(E_{\mathrm{F}}, B)$ are attributed to those of several predominant terms in (8.31) where the quantization rule

$$S(B) = \oint \boldsymbol{p} \cdot d\boldsymbol{r} = \oint \left(m_e \boldsymbol{v} - \frac{e}{c} \boldsymbol{A} \right) \cdot d\boldsymbol{r}$$

$$= \oint m_e \boldsymbol{v} \cdot d\boldsymbol{r} + \frac{e}{2\pi c} B\Theta = 2\pi\hbar N \qquad (8.32)$$

holds. In (8.32), $B\Theta$ is $2\pi\times$ the magnetic flux penetrating through the area enclosed by a periodic orbit (Θ is $2\pi\times$ area), and the integer N is defined in terms of the quantum number (n), winding number (w) and Maslov index (μ) as follows:

$$N = n + \frac{w}{2} + \frac{\mu}{4}.$$

In other words, with the use of the effective action

$$\widehat{S}(B) \equiv \frac{S(B)}{h} - \frac{w}{2} - \frac{\mu}{4} - \frac{1}{2}, \qquad (8.33)$$

the simple quantization rule $\widehat{S}(B) = n \, (= 1, 2, \cdots)$ provides the minima in resistivity $\rho_{xx}(E_{\mathrm{F}}, B)$. The relationship between resistivity and conductivity is $\rho_{xx} = \dfrac{\sigma_{xx}}{\sigma_{xx}^2 + \sigma_{xy}^2}$.

(In the present context, the argument on Hall conductivity is suppressed.)

The inset in Fig. 8.3(b) illustrates three short periodic orbits in the presence of the soft-wall antidot potentials, the orbit "a" surrounded by four antidots, the four-fold symmetric orbit "b" encircling a single dot, and the two-fold symmetric orbit "c" generated through a bifurcation of orbit "b". Let us concentrate on the phase factors in (8.31) and evaluate $\widehat{S}(B)$ using these short periodic orbits, and then we find $\rho_{xx}(E_{\mathrm{F}}, B)$ has minima at B values that make $\widehat{S}(B)$ integers, which is consistent with the experimental result described in Section 8.1. Note that $2n$ is related to ν in the high-field limit.

Strictly speaking, the present system is classically a mixed one where a given periodic orbit shows bifurcations into several children and $vice$ $versa$, for instance, by varying the B field. At each bifurcation point, some of the terms in (8.31) have divergent amplitudes due to the vanishing denominator, $|\det(\tilde{M}_{\gamma}^{p} - I)| = 0$ or $\mathrm{Tr}\tilde{M}_{\gamma}^{p} = 2$, which would give rise to the additional fluctuations of ρ_{xx}. In fact, Nihei et $al.$ observed the magneto-resistance ρ_{xx} of the triangular antidot lattice with a weak magnetic field. On the lower-field side of the fundamental commensurability peak, they discovered the anomalously large fluctuations besides the negative magneto-resistance induced by weak localization. This point will be investigated in Chapter 9.

On the other hand, the 2-d interacting electron gas in a strong magnetic field has also received wide attention. This system with a specific filling factor ν can be taken to be equivalent to the noninteracting electron gas composed of composite fermions (electrons accompanied by $\dfrac{1}{\nu}$ magnetic fluxes). These fluxes reduce the magnetic field effectively by $\overline{B} = \dfrac{1}{\nu}\Phi_0 n_e$ with the flux quantum Φ_0 and electron density n_e, and as a consequence the composite fermion feels a very weak field $B_{\mathrm{eff}} = B - \overline{B}$. The study of the nonlinear dynamics of noninteracting composite fermions under a B_{eff} field is progressing, and the analysis of the resistivity fluctuations in the strong field regime from the viewpoint of the weak B_{eff}-field commensurate orbits constitutes a challenging research field. Ando et $al.$[100] have developed independent numerical quantum-mechanical simulations of antidot lattices.

9

ORBIT BIFURCATIONS, ARNOLD DIFFUSION, AND COULOMB BLOCKADE

This chapter is devoted to other important and interesting themes bridging non-linear dynamics and quantum transport in quantum dots and antidots. First, the concept of orbit bifurcations is essential in nonlinear dynamics. Wherever orbit bifurcations occur, periodic orbits become degenerate and one should have recourse to an improved version of the Gutzwiller-type semiclassical trace formula and that of the semiclassical Kubo formula. We elucidate the significant role of orbit bifurcations in the anomalously large fluctuations observed in quantum transport in triangular antidot lattices. Second, we move on to investigate the semiclassical conductance for three-dimensional (3-d) ballistic open billiards. For partially or completely broken-ergodic 3-d billiards such as $SO(2)$ symmetric billiards, the dependence of the conductance on the Fermi wave number is dramatically changed by the lead orientation. Application of the symmetry-breaking magnetic field to such billiards brings about Arnold diffusion that cannot be seen in 2-d billiards. In contrast to the 2-d case, the enhanced negative magneto-resistance near the zero field occurs due to classical Arnold diffusion in addition to a possible quantal weak-localization correction. Finally we introduce a semiclassical theory on the statistical distribution for conductance-peak heights in the Coulomb blockade regime. In quantum dots with many electrons, the electron–electron correlation can be incorporated in the sense of mean-field theory and therefore one can construct a semiclassical theory of the tunneling current with use of a single-particle picture. The goal is to obtain the nonuniversal feature of the peak-height distribution and the peak-to-peak correlation which cannot be obtained by the standard random matrix theory.

9.1 Orbit Bifurcations in Triangular Antidot Lattices

The Gutzwiller-type semiclassical trace formula and the semiclassical Kubo formula work well only when there is no overlapping among saddle points, namely, when periodic orbits are isolated from each other. Otherwise, we should improve these original formulas. In fact, periodic orbits can show bifurcations when the system's parameters (e.g., magnetic or electric fields or billiard table shapes) are varied. At the point of bifurcations, periodic orbits become degenerate and improved semiclassical formulas should be invented.

As we have seen in Chapter 8, the antidot lattice is a realization of Sinai billiards, and it is a system of a high-mobility 2-d electron gas (2DEG) at a $GaAs/Al_xGa_{1-x}As$

145

heterostructure with imposed periodically arranged potential peaks (antidots). Since the mean free path l_e is much larger than the period of antidots d, and the Fermi wavelength is less than this period, the motion of electrons there can be considered as ballistic and semiclassical. The rich features of magneto-resistivity $\rho_{xx}(B)$ in square antidot lattices observed by Weiss *et al.* revealed the role of underlying classical periodic orbits including the commensurate orbits.

In the triangular antidot lattice, on the other hand, the NEC group (Nihei *et al.* [99]) observed anomalously large fluctuations in an experiment on magneto-resistivity $\rho_{xx}(B)$. These fluctuations, which lie on the lower-B field side of the fundamental commensurability peak, are too large to be interpreted on the basis of periodic orbit theory. In the beginning of this chapter, we suggest that this phenomenon can be attributed to orbit bifurcations. Within the semiclassical framework, we first analyze the conductivity σ_{xx} of fully chaotic triangular antidots in the low-to-intermediate magnetic field B. A combination of both the smooth classical part (evaluated by mean density of states) and the oscillation part (evaluated by periodic orbits) will prove only to bring about a monotonic decrease of $\rho_{xx}(B)$ with respect to the B field. On inclusion of the effect of orbit bifurcations due to the overlapping among typical shortest periodic orbits, however, several distinguished peaks of $\rho_{xx}(B)$ appear, which nicely explains the experimental facts (Ma and Nakamura[101]).

Nihei *et al.* fabricated triangular antidot lattices in the ballistic regime with the sample size ($\sim 15\,\mu$m) less than the mean free path ($l_e \sim 20\,\mu$m), and discovered the novel phenomena. In this experiment, the period (inter-antidot spacing) $d \sim 0.2\,\mu$m and the radius of each circular antidot (etched hole) $a \sim d/2$, and $\rho_{xx}(B)$ is expected to reflect properties of chaotic orbits in the presence of a low-to-intermediate magnetic field B less than 0.3 tesla. The experimental result displayed in Fig. 9.1(a) shows: (i) at each temperature (T), the magneto-resistivity $\rho_{xx}(B)$ on the average decreases monotonically with respect to B; (ii) it is however accompanied by three distinguished peaks at $B = 0.05, 0.105$ and 0.155 tesla, all of which are lying below the fundamental commensurability peak due to the cyclotron orbit encircling a single antidot; (iii) as the temperature is decreased, the magnitude of fluctuations is increased anomalously, that is, the peak heights amount to 20% of the mean resistivity at the lowest accessible temperature ($T = 0.07$K).

The above puzzling fluctuations seem to have a periodicity of a half of the flux quantum divided by the unit-cell area of the superlattice. But, in the B field regime under consideration the effect of orbital bending is essential, and it would not be acceptable to apply the concept of AAS oscillation due to quantum interference between a pair of time-reversal symmetric orbits (see Chapter 7). We investigate the above puzzling issue from the viewpoint of orbit bifurcations in nonlinear dynamics. In fact, in the B field regime where the cyclotron radius R_c is more or less larger than the interdot spacing d, the bent orbits can show successive bifurcations as shown below.

A schematic illustration of triangular antidots is given in Fig. 9.2. The Hamiltonian describing the electron dynamics reads

$$H = \frac{1}{2m_e}\left(\boldsymbol{p} + \frac{e}{c}\boldsymbol{A}(\boldsymbol{r})\right)^2 + U(\boldsymbol{r}),$$

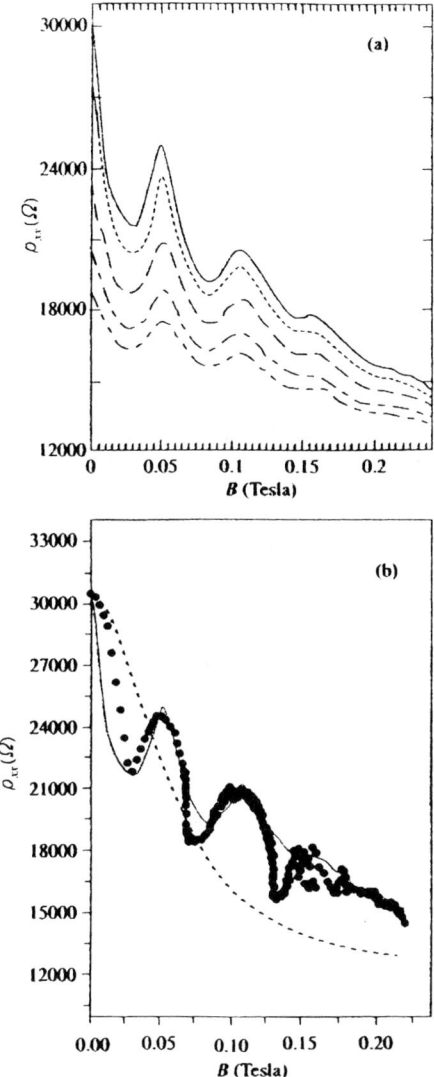

Fig. 9.1 (a) Magneto-resistivity $\rho_{xx}(B)$ of a triangular antidot lattice at different temperatures (T). Solid, dotted, dashed, dot-dashed, and two-dot-dashed lines correspond to $T = 0.07, 0.31, 0.6, 1.0$ and $1.4\,\mathrm{K}$, respectively (reproduced from[99]); (b) Computed theoretical results for $\rho_{xx}(B)$ at $T = 0.07\,\mathrm{K}$. The solid line is the experimental result. The dotted line is the theoretical result from (9.7) excluding the effect of bifurcations. The series of circular symbols is the theoretical result including the effect of bifurcations (reproduced from[101]).

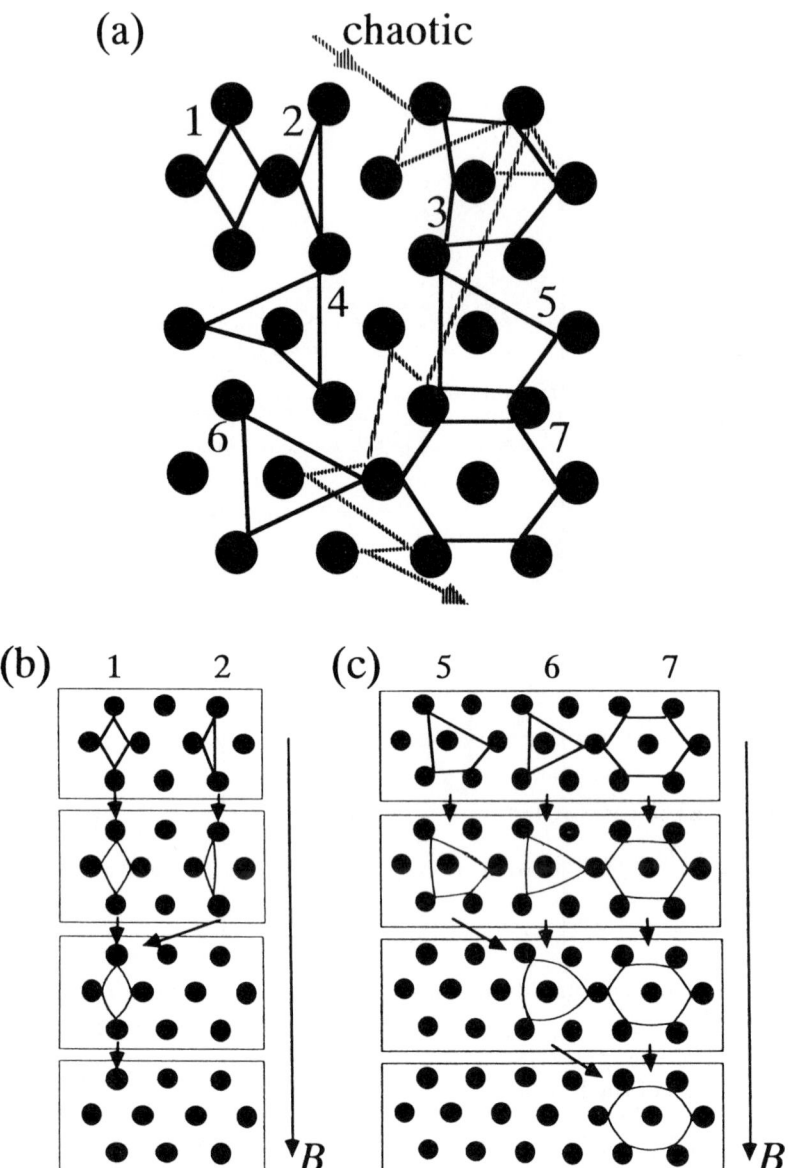

Fig. 9.2   (a) Schematic illustration of the triangular antidot lattice (filled circles) with typical periodic orbits (PO's) (solid lines) and chaotic orbit (broken line) at $B = 0$. While PO's 1, 2, 5-7 are completely confined by surrounding antidots, the confinement of PO's 3 and 4 is incomplete and polygons depicted by these orbits include corners with obtuse angle outside the confining region. (b) Tangent bifurcations induced by varying the B field. (c) Pitchfork bifurcations induced by varying the B field.

where $\boldsymbol{A}(\boldsymbol{r})$, $U(\boldsymbol{r})$, and m_e are the vector potential, the potential of triangular antidots, and the effective mass of the electron, respectively. For each of the antidots, we use the soft-wall potential

$$U(\mathbf{r}) = U(r) \equiv \begin{cases} E_{\mathrm{F}}(r/r_0 - 1 - s)^2 & (r \le r_0(1+s)) \\ 0 & (r > r_0(1+s)) \end{cases}$$

where r is the distance of the electron from the center of each antidot at $n d e_1 + m d e_2$ (with n, m integers and $e_1 = (1, 0)$ and $e_2 = (1/2, \sqrt{3}/2)$), E_{F} is the Fermi energy, and r_0 and s are other potential parameters.

9.1.1 Semiclassical Conductivity and Orbit Bifurcations

In a very strong field, electrons move only along the Landau orbit and the magneto-resistivity obviously shows Shubnikov–de Haas oscillation. On the contrary, we shall here concentrate on the low-to-intermediate field regime from zero to 0.3 tesla. Commensurate cyclotron orbits in this field regime, which have relatively large radii comparable to the sample size, cannot contribute to the quantum transport because of the lack of coherence and the finite temperature effect. So we conclude that the conductivity fluctuations of Nihei *et al.*[99] can be explained neither by the Shubnikov–de Haas oscillation nor by the commensurate cyclotron orbit theory. Noting the lattice periodicity $d \sim 0.2\,\mu\mathrm{m}$, it is sufficient to consider the deformation of periodic orbits (PO's) with their length much shorter than the sample size. In the semiclassical formalism, σ_{xx} for chaotic systems is given as a sum of the smooth part σ^0 and the oscillation part $\delta\sigma$ described by PO's:

$$\sigma_{xx}(E_{\mathrm{F}}, B) = \sigma^0_{xx}(E_{\mathrm{F}}, B) + \delta\sigma_{xx}(E_{\mathrm{F}}, B), \tag{9.1a}$$

$$\sigma^0_{xx}(E_{\mathrm{F}}, B) = g_s e^2 g(E_{\mathrm{F}}) \int_0^\infty dt < v_x(t)v_x(0) >_{\boldsymbol{p},\boldsymbol{r}} \exp(-t/\tau), \tag{9.1b}$$

$$\delta\sigma_{xx}(E_{\mathrm{F}}, B) = \frac{2g_s}{\Omega}\frac{e^2}{h}\sum_{po} C^{po}_{xx} \frac{\Xi(L_{po}/L_T)\exp\left(-\frac{L_{po}}{2\tau v_{\mathrm{F}}}\right)}{|\det(\tilde{M}_{po} - I)|^{1/2}}$$
$$\times \cos(S_{po}/\hbar - \mu_{po}\pi/2), \tag{9.1c}$$

where $g(E_{\mathrm{F}})$ is the mean density of states at E_{F}, and $< \cdots >_{\boldsymbol{p},\boldsymbol{r}}$ implies the phase space average. $g_s(= 2), \Omega, L_{po}, \tau, v_{\mathrm{F}}, S_{po}, \mu_{po}$, and $\tilde{M}_{po}$ are the spin factor, the area of the unit cell of the system, the length of the PO, the scattering time, the Fermi velocity, the action of the PO, the Maslov index and the stability matrix, respectively. $\Xi(L_{po}/L_T) = \dfrac{L_{po}/L_T}{\sinh(L_{po}/L_T)}$ with the cutoff length $L_T = v_{\mathrm{F}}\hbar/(\pi k_{\mathrm{B}}T)$ is a thermal dampimg factor. The velocity-correlation function reads

$$C^{po}_{xx} = \int_0^{\tau_{po}} dt \int_0^\infty dt' v_x(t)v_x(t + t') \exp(-t'/\tau_{po}) \tag{9.2}$$

with τ_{po} the period of the PO. It should be pointed out that (9.1c) is only valid for completely chaotic systems in which all PO's are isolated.

Each PO term in the oscillating part $\delta\sigma_{xx}$ in (9.1c) includes the denominator $|\det(\tilde{M}_{po} - I)|^{1/2}$. This factor has appeared as a result of the stationary-phase approximation and represents the curvature of the action at each saddle point (periodic orbit) that are mutually isolated. In the case that a pair of adjacent saddle points merges, the curvature vanishes, namely,

$$\det(\tilde{M}_{po} - I) = 0. \tag{9.3}$$

Then the contribution to $\delta\sigma_{xx}$ from this degenerate point becomes divergent! Noting $\det(\tilde{M}_{po}) = 1$ due to the conservative nature of Hamiltonian dynamics, the equality (9.3) is equivalent to $\mathrm{Tr}\tilde{M}_{po} = 2$ for the systems in two dimensions. The merging of PO's, which is called the orbit bifurcation, widely occurs in generic or mixed systems such as those with soft-walled potentials and/or with the applied magnetic field, when the systems' parameter crosses some critical value.

An example of orbit bifurcation is depicted in Fig. 9.2(b). Periodic orbits PO 1 trapped by four antidots and PO 2 trapped by three antidots, are mutually different at the magnetic field $B = 0$. By increasing the B field, both PO's 1 and 2 begin to be bent and become degenerate at some field $B = B_c$. At $B > B_c$, both PO's 1 and 2 vanish, making way for a *ghost orbit*. Conversely, by decreasing the B field from above $B = B_c$, the ghost orbit bifurcates to PO's 1 and 2. This kind of bifurcation is called the **tangent bifurcation**. In the neighborhood of the bifurcation point, the phase space of the underlying classical dynamics has a local structure consisting of two kinds of saddle points, stable and unstable (see Fig. 9.3(a)). On the other hand, Fig. 9.2(c) shows that, with *decrease* of B from the high field, PO 7 trapped by six antidots generates PO 6 trapped by three antidots at $B = B_{c2}$ and PO's 6 and 7 coexist at $B < B_{c2}$. Further decrease of B lets PO 6 generate PO 5 trapped by four antidots at $B = B_{c1}$, and PO's 5–7 coexist at $B < B_{c1}$. This is the **pitchfork bifurcation**. The corresponding phase space structure is given in Fig. 9.3(b). Since $\mathrm{Tr}\tilde{M}_{po} = 2$ at all these bifurcation points, one needs a new formula that replaces (9.1c).

As explained in Chapter 8, $\delta\sigma_{xx}$ in (9.1c) was originally derived by the stationary-phase approximation for double integrals in the semiclassical path-integral representation of the Kubo formula. The procedure was based on: (1) the assumption that all PO's are isolated; (2) approximation of the quantal fluctuations around each PO by the quadratic form; and (3)application of the formula for the Gauss–Fresnel integrals. For the purpose of incorporating the nonlocal structure of fluctuations in the stationary-phase approximation, we first make the Legendre transformation of the action function to avoid the problem of divergence due to the degeneracy. The action function is then represented by the initial momentum $\boldsymbol{p}$ and final coordinate $\boldsymbol{q}'$, and is given by[56-58]

$$S(\boldsymbol{q}', \boldsymbol{p}, E) = S_0(E) + q'p/2 - \eta p^2 - V(q') \tag{9.4}$$

with $\eta = \pm 1$. $V(q')$ is the **nonlocal potential** describing the nonlocal structure of fluctuations. Here one employs the uniform approximation[4], that is, one provides the approximate function for $V(q')$ so that, far away from the bifurcation point, the simple PO sum in (9.1c) can be recovered. $V(q')$ is a third order polynomial in the

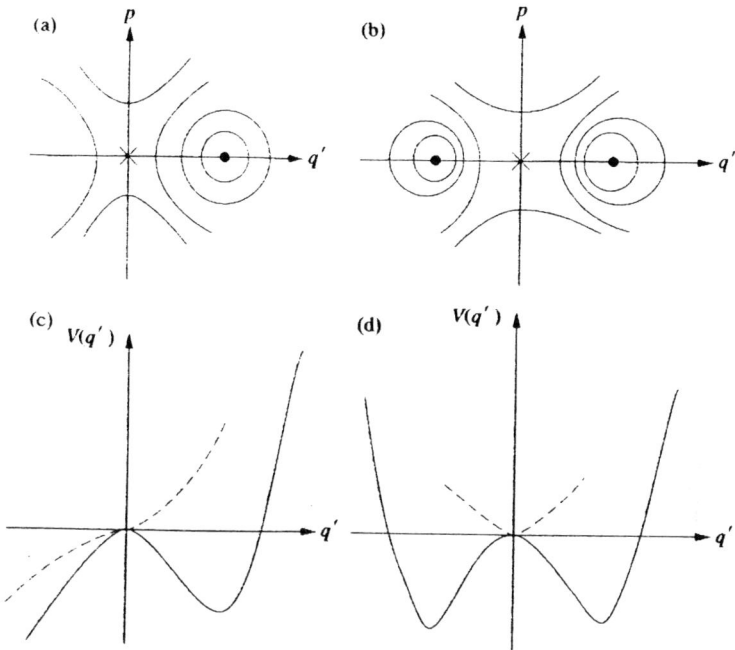

Fig. 9.3 Phase space structure and nonlocal potential: (a), (c) tangent bifurcation; (b), (d) pitchfork bifurcation. (a), (b) describe structures after the bifurcation, and • and × are stable and unstable fixed points, respectively. Solid and broken curves in (c), (d) are potentials after and before the bifurcation, respectively.

case of the tangent bifucation (Fig. 9.3(c)), and a fourth order polynomial in the case of the pitchfork bifucation (Fig. 9.3(d)).

The integration over q' can be done with use of the **diffraction-catastrophe integral**. The resulting improved formulas are expressed in terms of the Airy function[52]. Readers should have recourse to the original references of Sieber[56,57] and Schomerus and Sieber[58]. Interestingly, however, the quantum correction at the bifurcation point is $\delta\sigma_{xx} \sim \hbar^{-7/6}$ for the tangent bifurcation, while it is $\delta\sigma_{xx} \sim \hbar^{-5/4}$ for the pitchfork bifurcation. In the semiclassical regime ($\hbar \to 0$), these contributions are much larger than the conventional result $\delta\sigma_{xx} \sim \hbar^{-1}$. The improved formula will play a powerful role in the following.

9.1.2 Quantum Correction without Orbit Bifurcations

First, we shall evaluate the quantum correction without taking into consideration the effect of orbit bifurcations. The triangular antidot lattice under consideration is completely chaotic in the B field less than 0.3 tesla, i.e., in the low-to-intermediate B

field. In other words, the measure of chaotic orbit is unity (100%) and does not change with the B field. Then the smooth part σ^0_{xx} is determined by chaotic scattering orbits, which we shall calculate first. In (9.1b), it is enough to do the integration from $t = 0$ to the time of the first collision, because after the collision the electron will lose its memory and give no contribution to the integration. So the integration in (9.1b) will be reduced to

$$v_F^2 \int_0^\infty dt < v_x^2(0)/v_F^2 >_{p,r} \exp(-t/\tilde{t}) \exp(-t/\tau) = \frac{1}{2} v_F^2/(1/\tilde{t} + 1/\tau),$$

where $\exp(-t/\tilde{t})$ is the probability for the electron to show a free flight up to the first collision time $\tilde{t} \sim d/v_F$ and the averaging of $< v_x^2(0)/v_F^2 >_{p,r}$ over the direction gives a factor $1/2$. Although in (9.1b) $g(E_F) \sim 1/h^2$, other factors are very small, so we will see the magnitude of σ^0_{xx} is comparable to that of $\delta\sigma_{xx}$. Concentrating on the experimental values at $T = 0.07$ K, we obtain $\sigma^0_{xx} = 1.94 \times 10^{-4} \Omega^{-1}$. Subtracting σ^0_{xx} from the experimental value $\sigma_{xx}(B = 0) \approx 1/\rho_{xx}(B = 0)$ obtained from Fig. 9.1, we can evaluate $\delta\sigma_{xx}(B = 0) = -1.61 \times 10^{-4}$ (**negative value!**).

The B field does not enter into the mean density of states. Moreover, the system remains fully chaotic, keeping the portion of scattering orbits unchanged with respect to the B field up to $B = 0.3$ tesla. Therefore σ^0_{xx} should be independent of the B field and we should proceed to investigate the influence of the B field on $\delta\sigma_{xx}$ by using the semiclassical PO theory.

To simplify the analysis, we first investigate the B dependence of the averaged magnitude of $\delta\sigma_{xx}$ in (9.1c) in the diagonal approximation, and then the precise structure of fluctuations. In the experimental temperature region $T \gg \hbar^2/(2m_e L^2 k_B) \sim 0.01$ K with L the characteristic length of the system, the thermal fluctuations smear out fine structures on the energy scale of mean level spacing, and the asymptotic form $\Xi(L_{po}/L_T) \approx 2(L_{po}/L_T) \exp(-L_{po}/L_T)$ can be employed. Under these circumstances, the variance $\langle \delta\sigma_{xx}^2 \rangle$ can be divided into two parts

$$\langle \delta\sigma_{xx}^2 \rangle = \overline{\delta\sigma_{xx}^2} + \langle \delta\sigma_{xx}'^2 \rangle$$

with the B-independent part $\overline{\delta\sigma_{xx}^2} = \langle \delta\sigma_{xx}'^2(B = 0) \rangle = \frac{1}{2}\delta\sigma_{xx}^2(B = 0)$ and the B-dependent part $\langle \delta\sigma_{xx}'^2 \rangle$ given by

$$\langle \delta\sigma_{xx}'^2 \rangle = 2C \sum_{po} \left(\frac{L_{po}}{L_T} \right)^2 \exp\left\{ -\left(2\frac{L_{po}}{L_T} + \frac{L_{po}}{\tau v_F} \right) \right\}$$

$$\times \frac{\cos^2(\frac{1}{\hbar} \oint_{po} \boldsymbol{p} \cdot d\boldsymbol{r} - \mu_{po}\pi/2)}{|\det(\tilde{M}_{po} - I)|} \cos(2eB\Theta_{po}/(\hbar c))$$

$$(9.5)$$

with $C = \dfrac{16e^4 \bar{C}_{xx}^2}{(h\Omega)^2}$. In (9.5), Θ_{po} is the area enclosed by the PO (note that the multiplication factor 2π is not included in Θ_{po} throughout this chapter), and, noting that the integration over t in (9.2) would yield almost the same value for different PO's, the averaged value $\bar{C}_{xx}^2$ is used to replace $C_{xx}^{po\,2}$.

So long as there is no degeneracy among PO's, (9.1c) is justified and one can apply the Hannay and Ozorio de Almeida (H-OdA) sum rule for PO's,

$$\sum_{po} 1/|\det(\tilde{M}_{po} - I)| \cdots = \int_0^\infty dL/L \cdots .$$

In (9.5), $\cos^2(\cdots) = 1/2 + \cos(2 \times \cdots)/2$, but the H-OdA sum rule lets

$$\cos\left(\frac{2}{\hbar} \oint_{po} \boldsymbol{p} \cdot d\boldsymbol{r} - \mu_{po}\pi\right)$$

terms vanish since even and odd Maslov numbers cancel these terms with each other, and only the average part survives.

We now evaluate (9.5) by having recourse to the periodic orbit distribution: substituting into (9.5) the probability distribution for the area Θ enclosed by PO with a given length L (see (3.51) in Chapter 3),

$$P_L(\Theta) = \frac{1}{\sqrt{2\pi L\alpha_L}} \exp\left(-\frac{\Theta^2}{2L\alpha_L}\right),$$

and, integrating over Θ and L, we have

$$\langle \delta\sigma_{xx}'^2 \rangle = C \int_0^\infty dL \int_{-\infty}^\infty d\Theta \frac{1}{L} \left(\frac{L}{L_T}\right)^2$$
$$\times \exp\left\{-\left(2\frac{L}{L_T} + \frac{L}{\tau v_F}\right)\right\} \cos(2eB\Theta/(\hbar c)) P_L(\Theta)$$
$$= C \left(2 + \frac{L_T}{\tau v_F} + \frac{2e^2\alpha_L L_T}{\hbar^2} B^2\right)^{-2}, \tag{9.6}$$

arriving at the diagonal approximation for the oscillation part of the conductivity:

$$\delta\sigma_{xx}(B) = \delta\sigma_{xx}(B=0)\sqrt{1/2 + \langle \delta\sigma_{xx}'^2 \rangle / \delta\sigma_{xx}^2(B=0)}. \tag{9.7}$$

Using (9.7) together with σ_{xx}^0 and the **negative value** $\delta\sigma_{xx}(B=0)$ derived already, we evaluate the resistivity

$$\rho_{xx} = (\sigma_{xx}^0 + \delta\sigma_{xx}(B))^{-1}$$

at $T = 0.07\,\mathrm{K}$, yielding the result plotted as the dotted line in Fig. 9.1(b), where the monotonic decrease of resistivity predicted from (9.6) and (9.7) fits the experimental result well except at several distinct peaks. Equations (9.6) and (9.7) also show that the higher the temperature, the more slowly ρ_{xx} decreases, the same feature as found in Fig. 9.1(a). However, this does not as yet explain the anomalously large fluctuations in the experiment. Even if the off-diagonal terms were incorporated in $\delta\sigma_{xx}(B)$, the evaluated resistivity fluctuations are too small to reproduce the experimentally observed fluctuations that amount to 20% of the mean resistivity.

9.1.3 Orbit Bifurcations and Anomalous Resistivity Fluctuations

As mentioned repeatedly, (9.1c) is valid only when all the PO's are isolated. In the low-to-intermediate field regime, the B field will bend the PO's and some PO's will overlap each other at some special values of B, where the bifurcation of the orbit occurs. At these points $\mathrm{Tr}\tilde{M}_{po} = 2$ and the amplitude factor in (9.1c) becomes divergent. At bifurcations, the uniform approximation mentioned before should be employed. From (9.1c), one finds that only the bifurcation of several shortest orbits is needed since long orbits are suppressed by the finite temperature effect. To judge whether the bifurcation occurs, we should compute the 4×4 stability matrix M_{po} for PO's because $\mathrm{Tr}M_{po} = \mathrm{Tr}\tilde{M}_{po} + 2 = 4$ at bifurcations. The definition for M_{po} reads $M_{po} = M(t = T_{po})$ which characterizes the growth of the variation $\delta\gamma'(t) = M\delta\gamma(t = 0)$; the equation of motion for M is

$$\frac{dM}{dt} = J \frac{\partial^2 H}{\partial \gamma^2}\bigg|_{\gamma_{po}(t)} M, \qquad (9.8)$$

where

$$J = \begin{pmatrix} 0 & I \\ -I & 0 \end{pmatrix}$$

with I the 2×2 unit matrix and γ denotes a pair of conjugate coordinates, $(r, p - \frac{e}{c}A)$. The initial M is a unit matrix since the unit displacement is conceived at $t = 0$.

From Fig. 9.2, one finds that PO 2 is bent by the magnetic field and begins to overlap with PO 1, and a similar phenomenon occurs between PO's 3 and 4. When the symmetry of PO 5 becomes higher, it overlaps with PO 6. Furthermore, PO 6 changes into PO 7 at a certain special value of B. Taking the potential parameters $r_0 = 0.093\,\mu\mathrm{m}$ and $s = 0.35$, we compute $\mathrm{Tr}M$ for these PO's and display the results in Fig. 9.4. From Fig. 9.4(a), we find that PO's 1 and 2 merge with each other and simultaneously disappear at $B = 0.057\,\mathrm{T}$ [tesla], and in Fig. 9.4(b), so do PO's 3 and 4 at $B = 0.109\,\mathrm{T}$. Therefore they show tangent bifurcations. Moreover, PO's 5 and 6 in Fig. 9.4(c) merge with each other and only PO 6 survives at $B = 0.144\,\mathrm{T}$; subsequently PO 6 meets PO 7 at $B = 0.152\,\mathrm{T}$ and only PO 7 survives. So they show two successive pitchfork bifurcations. Precisely speaking, each PO shows two kinds of bending under the B field. One is its inflation toward the outside, and the other is its contraction toward the inside. Their directions of electronic motion are opposite. Here only the inflation orbits play an important role in the bifurcation of orbits and the contraction orbits, which are very stable, are irrelevant. Computing all necessary variables and using the uniform approximation, we can derive the new oscillation part $\delta\sigma_{xx}$ at these bifurcations as

$$\delta\sigma_{xx} \sim \frac{e^2}{h^{7/6}}\Xi(L_{po}/L_T)C_{xx}^{po}\cos(S_{po}/\hbar - \mu_{po}\pi/2) \qquad (9.9)$$

at the tangent bifurcations and

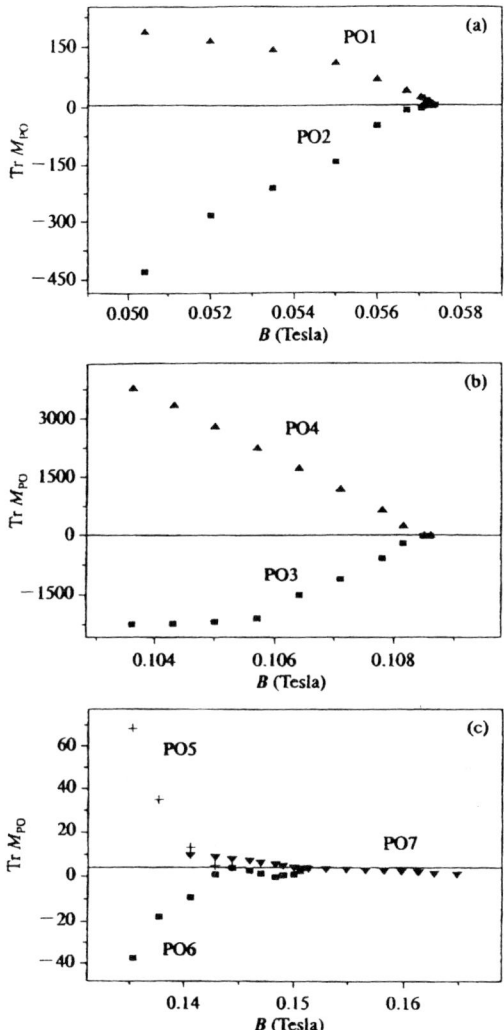

Fig. 9.4   Values of TrM: (a) PO's 1 (triangle) and 2 (square); (b) PO's 3 (square) and 4 (triangle); (c) PO's 5 (cross), 6 (square) and 7 (downward triangle). The horizontal line represents Tr$M = 4$. The magnitude of TrM is very much affected by the geometric nature of the PO, and the anomalously large scale of the ordinate in Fig. 9.4(b) is due to the incomplete confinement of PO's 3 and 4 (reproduced from[101]).

$$\delta\sigma_{xx} \sim \frac{e^2}{\hbar^{5/4}} \Xi(L_{po}/L_T) C_{xx}^{po} \cos(S_{po}/\hbar - \mu_{po}\pi/2) \qquad (9.10)$$

at the pitchfork bifurcations, which are much larger than the ordinary contributions of $O(1/\hbar)$ from each isolated PO in (9.1c).

Incorporating the interference terms leading to the above anomalous corrections, the calculated results for the resistivity $\rho_{xx} = 1/(\sigma_{xx}^0 + \delta\sigma_{xx}(B))$ are plotted in a series of circle symbols in Fig. 9.1(b), where the minima of $\delta\sigma_{xx}$, which are responsible for the bifurcation points, correspond to the three distinguished peaks in ρ_{xx} observed in Nihei *et al.*'s experiment. Obviously, the lengths of PO's 1 and 2 at the bifurcation point are the shortest. Since the stronger magnetic field leads to the larger inflation of the PO, the orbits at the bifurcation in Fig. 9.4(c) are the longest and the ones in Fig. 9.4(b) are intermediate. Bifurcation of other longer PO's is suppressed due to the finite temperature effect. We see that the first peak at $B = 0.05$ T is the highest, the one at 0.10 T is the second highest, and the peak at 0.15 T is the lowest. One can also predict that the increased temperature suppresses the height of these peaks. Both the locations and anomalously large heights of all these peaks are in very good agreement with the experimental result in Fig. 9.1(a).

It should be noted that (i) a deformation of the potential for each of the antidots does not affect the positions of bifurcations because these positions are determined by the symmetry of the PO's in Fig. 9.2; and (ii) the anomalously large oscillations in resistivity analyzed here may be observed only in the ballistic regime. The observed small oscillation in resistivity in diffusive square antidot lattices was attributed to the change of the level density of systems with respect to the B field, which corresponds to the change of phase factors in nondiagonal terms beyond (9.5).

Thus, quantum transport in antidot lattices provides the stages where one can capture the quantal signature of orbit bifurcations.

9.2 Arnold Diffusion and Negative Magneto-Resistance

Let us move on to the second theme in this chapter. Since the ground-breaking experiments on the magneto-conductance for submicron-scale stadium and circle billiards with conducting leads, growing attention has been paid to ballistic quantum transport in two-dimensional (2-d) quantum dots and its relationship with the underlying nonlinear dynamics. Experimentally, the dependence of the conductance was explored for the lead orientation as well as for the Fermi wave number and magnetic field. Theoretically, the semiclassical conductance formula was derived for 2-d billiards by combining the semiclassical Green function with the Landauer formula.

The salient phenomenon in this field is the negative magneto-resistance, namely, the decrease of resistance on application of the weak magnetic field B, which was originally evidenced in the numerical quantal analysis of the conductance for ballistic 2-d quantum dots. This phenomenon at first looks curious because the B field induces the cyclotron motion of electrons which should suppress the electronic transmission. The semiclassical theory (based on the saddle-point approximation for the Feynman path-integral representation of the Green function) can also lead to a transmission coefficient (dimensionless conductance) consistent with the quantal result in the large wave number region. Using the semiclassical reflection coefficient, as explained in Chapter 7, Baranger et $al.$[46,47] pointed out that the negative magneto-resistance in ballistic 2-d billiards should be ascribed to weak localization based on the quantum interference between a pair of time-reversal symmetric back-scattering orbits. Precisely speaking, the semiclassical theory is qualitatively correct: although the semiclassical transmission coefficient does not clearly yield the weak-localization correction without inclusion of the small-angle-induced diffraction effect (Aleiner and Larkin[80], Takane and Nakamura[84]), it can qualitatively reproduce the quantal negative magneto-resistance.

In the following we shall extend the semiclassical conductance formula for 2-d billiards to that for three-dimensional (3-d) billiards and investigate the **quantum transport in open 3-d billiards** from the viewpoint of **Arnold diffusion** (Ma et $al.$[104]). Arnold diffusion – a key concept in high-dimensional nonlinear dynamical systems – is one of the anomalous diffusions expected in ballistic systems with more than two degrees of freedom, namely, in 3-d systems. While this diffusion is a classical phenomenon, it should have an effect on quantum transport, which can only be revealed by using the semiclassical theory. As for the realizability of three-dimensional (3-d) billiards, ballistic 3-d quantum dots will be fabricated, for example, by exploiting drying etching processess with focused-ion beams applied to $Al_x Ga_{1-x}$ $As/GaAs/Al_x Ga_{1-x} As$ double heterostructures, which will be less than both the elastic mean free path and phase coherence length. Studies of equilibrium properties of the closed 3-d billiards are already started, and the quantum transport in open 3-d billiards constitutes a novel subject.

9.2.1 Semiclassical Conductance for Open Three-Dimensional Billiards

The conductance of open 3-d billiards connected to a pair of rectangular parallel-piped lead wires can be expressed by the transmission coefficient T as $g = \dfrac{2e^2}{h} T =$

$\frac{2e^2}{h} \sum_{n,m=1} |t_{nm}|^2$, where t_{nm} is the S matrix element between the incoming mode
$\mathbf{m} = (m_x, m_y)$ and the outgoing mode $\mathbf{n} = (n_x, n_y)$, and the double summation is
taken over all propagating modes. By extending the formula for 2-d open billiards,
the transmission amplitude connecting those modes at the Fermi energy E_F reads as

$$t_{nm} = -i\hbar\sqrt{v_n v_m} \int dx'' dy'' dx' dy' \psi_n^*(x'', y'')$$
$$\times G(x'', y'', z''; x', y', z'; E_F)\psi_m(x', y'), \tag{9.11}$$

where v_m and v_n are the longitudinal velocities of electrons for the incoming and outgoing modes, respectively, and x', y', z' and x'', y'', z'' are the corresponding *local* coordinates with the transverse (x, y) and longitudinal (z) components; the positive direction for $z'(z'')$ is inward (outward) the billiard system cavity. $G(x'', y'', z''; x', y', z'; E_F)$
is the Green function for an electron propagating from the entrance to the exit.
$\psi_m(x', y')$ and $\psi_n(x'', y'')$ are the transverse components of the wave functions at
the leads. To simplify the problem, the cross-section of the leads is assumed to
be a square with side length l. Then $\psi_m(x', y') = \frac{2}{l}\sin(m_x\pi x'/l)\sin(m_y\pi y'/l)$ and
$\psi_n(x'', y'') = \frac{2}{l}\sin(n_x\pi x''/l)\sin(n_x\pi y''/l)$. In the ballistic regime, the Green function
from the entrance $(\mathbf{r}')$ to the exit $(\mathbf{r}'')$ can be replaced by the Gutzwiller semiclassical
path-integral expression

$$G_{scl}(x'', y'', z''; x', y', z'; E_F) = \frac{2\pi}{(2\pi i\hbar)^2 v_F} \sum_{s(\mathbf{r}', \mathbf{r}'')} \left| \det\left(\frac{\partial \mathbf{p}_\perp'}{\partial \mathbf{r}_\perp''}\right) \right|^{1/2}$$
$$\times \exp\left(\frac{i}{\hbar}S_s(\mathbf{r}'', \mathbf{r}', E_F) - i\pi\mu_s/2\right), \tag{9.12}$$

where S_s and μ_s are the action and Maslov index for a trajectory s, and v_F is the
Fermi velocity. Here, to avoid direct transmission without any bouncing at the billiard
wall we may place a stopper inside the billiard system.

In the integration (9.11), we introduce the entrance angle θ' between the initial
momentum vector of the orbit and the z' axis, and the angle φ' between the projection
of this momentum onto the x'-y' plane and the x' axis. Angles θ'' and φ'' are defined
for the exit in the same way. Then, we can rewrite $\left| \det\left(\frac{\partial \mathbf{p}_\perp'}{\partial \mathbf{r}_\perp''}\right) \right|$ for each trajectory
s as $\left| \det\left(\frac{\partial(p_\perp^{1'}, p_\perp^{2'})}{\partial(r_\perp^{1''}, r_\perp^{2''})}\right) \right| = p_F^2 A_s$, where $A_s = \left| \frac{\sin\theta'}{\cos\theta''}\left(\frac{\partial\varphi'}{\partial x''}\frac{\partial\theta'}{\partial y''} - \frac{\partial\varphi'}{\partial y''}\frac{\partial\theta'}{\partial x''}\right) \right|$. For
large mode numbers, the integration over the transverse coordinates in (9.11) can
be performed in the stationary-phase approximation, and the transmission amplitude
becomes

$$t_{nm} = -i\hbar\frac{16\pi}{l^2 p_F} \sum_{s(\bar{m}_x, \bar{m}_y, \bar{n}_x, \bar{n}_y)} \text{sgn}(\bar{m}_x)\text{sgn}(\bar{m}_y)\text{sgn}(\bar{n}_x)\text{sgn}(\bar{n}_y)$$

$$\times (\cos\theta' \cos\theta'')^{1/2} (A_s/B_s)^{1/2} \exp\left(\frac{i}{\hbar}\tilde{S}_s - i\tilde{\mu}_s \frac{\pi}{2}\right) \qquad (9.13)$$

with

$$B_s = \left| \sin 2\varphi' \left(\frac{\partial\theta'}{\partial y'}\frac{\partial\theta'}{\partial x'}\cos^2\theta' - \frac{\partial\varphi'}{\partial y'}\frac{\partial\varphi'}{\partial x'}\sin^2\theta'\right) \right.$$
$$\left. + \sin 2\theta' \left(\frac{\partial\varphi'}{\partial y'}\frac{\partial\theta'}{\partial x'}\cos^2\varphi' - \frac{\partial\theta'}{\partial y'}\frac{\partial\varphi'}{\partial x'}\sin^2\varphi'\right) \right|$$
$$\times \left| ' \to '' \right|, \qquad (9.14)$$

where $\bar{m}_x = \pm m_x, \bar{m}_y = \pm m_y$, $\bar{n}_x = \pm n_x, \bar{n}_y = \pm n_y$, and $\tilde{S}_s = S(\mathbf{r}''_s, \mathbf{r}'_s, E_F) + \hbar\pi\bar{m}_x x'_s/l + \hbar\pi\bar{m}_y y'_s/l - \hbar\pi\bar{n}_x x''_s/l - \hbar\pi\bar{n}_y y''_s/l$, and $\tilde{\mu}$ is the Maslov index which includes an extra phase coming from the possible sign change in each of the four Fresnel integrals. The above result means that only those isolated trajectories connecting a pair of transverse planes at discrete angles $\sin\theta' \cos\varphi' = \bar{m}_x\pi/k_F l$, $\sin\theta' \sin\varphi' = \bar{m}_y\pi/k_F l$, $\sin\theta'' \cos\varphi'' = \bar{n}_x\pi/k_F l$, and $\sin\theta'' \sin\varphi'' = \bar{n}_y\pi/k_F l$ dominate the conductance.

Using (9.13), we shall first analyze a completely chaotic 3-d billiard system with a pair of conducting leads. In this case, the electron injected from the incoming lead will bounce at the billiard boundary in an ergodic way before reaching the exit, leading to amplitudes of the same order for both transmission and reflection coefficients. In the large mode number case, the summations over modes are replaced by integrations and the transmission coefficient $T = \sum_{\mathbf{n,m}} |t_{\mathbf{nm}}|^2$ is rewritten as (Ma et $al.$[104])

$$T \approx \frac{256 k_F^2}{\pi^2} \int_{-1}^{1} \sin\theta' d(\sin\theta') \int_{-\pi}^{\pi} d\varphi' \int_{-1}^{1} \sin\theta'' d(\sin\theta'') \int_{-\pi}^{\pi} d\varphi'' \sum_{s,u} \cos\theta' \cos\theta''$$
$$\times \left(\frac{A_s A_u}{B_s B_u}\right)^{\frac{1}{2}} \exp\left(\frac{i}{\hbar}(\tilde{S}_s - \tilde{S}_u) - i\frac{\pi}{2}(\tilde{\mu}_s - \tilde{\mu}_u)\right). \qquad (9.15)$$

Noting the dimension of $\left(\dfrac{A_s A_u}{B_s B_u}\right)^{1/2}$ to be l^2 and using the diagonal approximation, the above integration can be evaluated as $T \propto (k_F l)^2$. The conductance for 3-d chaotic billiards is thus given by $g \propto \dfrac{e^2}{h}(k_F l)^2$ without any dependence on the lead orientation.

9.2.2 Completely or Partially Broken-Ergodic 3-d Billiards

However, the situation will be dramatically changed for the case of completely or partially broken-ergodic 3-d billiards. As examples, we choose $SO(2)$-symmetric billiards. Completely integrable (broken-ergodic) 3-d billiards are available by rotating, e.g., the 2-d elliptic billiards (on the Y-Z plane) around the Z axis (see Fig. 9.5).

Note that X, Y, Z are the *global* coordinates. Partially broken-ergodic 3-d billiards are also available in a similar way by rotating, e.g., the 2-d stadium. All these $SO(2)$-symmetric 3-d billiards have angular momentum L_Z as a constant of the motion.

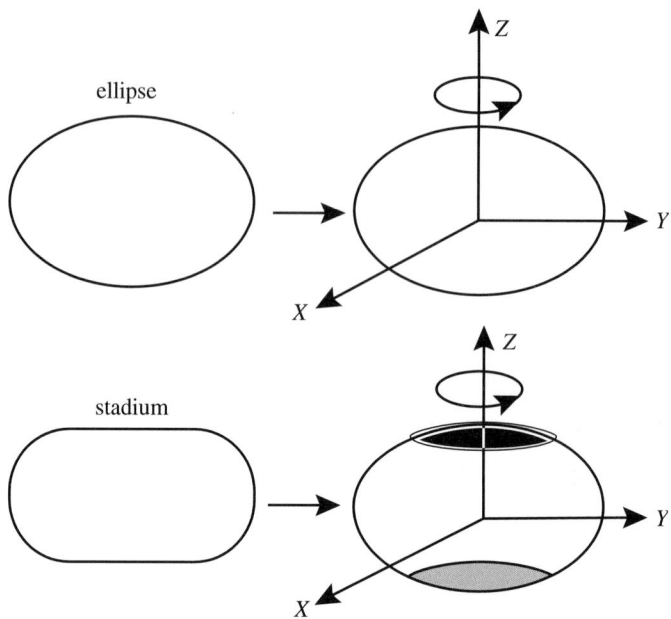

Fig. 9.5   Schematic illustration of $SO(2)$-symmetric 3-d billiards. The 2-d billiards are rotated around the Z axis to generate 3-d billiards.

In an open system version of each of these billiards, one may place inside a smaller stopper of similar shape to prevent the direct transmission of electrons. The resultant 3-d *shell* billiards still retain the $SO(2)$ symmetry.

First, consider the perpendicular case (i) when the incoming and outgoing leads are connected with the billiard perpendicularly at points A and C, respectively (see Fig. 9.6(a)). To reach the exit C, electrons should have vanishing angular momentum, $L_Z = 0$. Therefore, only the electrons with the initial velocity vector at the entrance A lying in the Y-Z plane, which have $L_Z = 0$, can reach C because of the conservation of L_Z; other incoming electrons falling into the trajectories out of this plane have $L_Z \neq 0$, and they cannot reach C and should return to A. In deriving the transmission coefficient in the continuum-mode approximation for this lead orientation, the integration prior to (9.15) is reduced to the line integration over θ' and θ'' only, yielding the conductance $g \propto \frac{e^2}{\hbar} k_F l$ (linear in k_F!) as in the case of 1-d quantum transport. (The reflection coefficient, which is obtained by applying the same procedure as in chaotic billiards, is proportional to $(k_F l)^2$.) We shall then consider the parallel case (ii) when the incoming and outgoing leads are placed parallel at points Q and C, respectively (see Fig. 9.6(b)). Any incoming electron from the entrance Q has vanishing angular momentum, $L_Z = 0$, and therefore satisfies the condition to reach the exit C. For this lead orientation, each electron can reach C or return to Q with probability

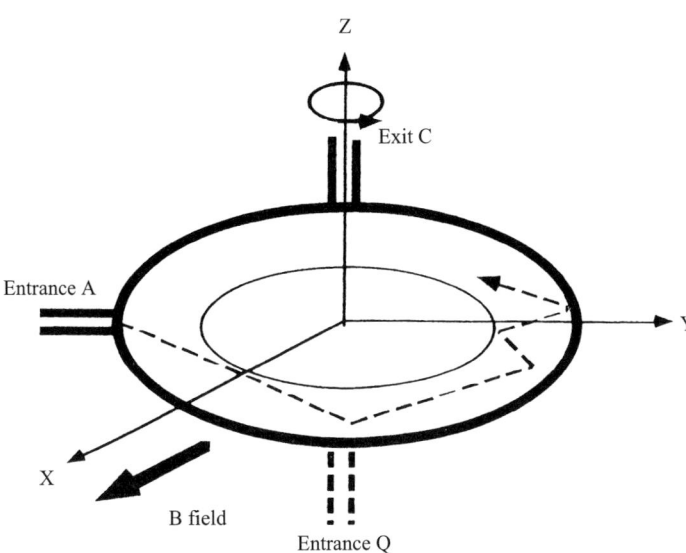

Fig. 9.6 $SO(2)$-symmetric 3-d billiard and lead alignments with common exit C: perpendicular case with entrance A and parallel case with entrance Q. Dual 2-d ellipses sharing the same foci in the Y-Z plane are rotated around the Z axis to generate 3-d elliptic shell billiards. The longer radii of outer and inner ellipses are taken as $2\Re$ and $1.85\Re$, respectively. The shorter radius of the outer ellipse is $0.8\Re$. The side length l for the square cross-section of the lead is $\Re/20$.

of the same order, which leads to $g \propto \dfrac{e^2}{h}(k_{\mathrm{F}}l)^2$.

Generally, in completely or partially broken-ergodic 3-d billiards, we have $g \propto \dfrac{e^2}{h}(k_{\mathrm{F}}l)^\gamma$ with $1 \leq \gamma \leq 2$, and the exponent γ is determined by how the billiard symmetry is broken by the lead orientation. It should be noted that open 2-d billiards always share $g \propto \dfrac{e^2}{h}k_{\mathrm{F}}l$, independently of both integrability and lead orientations.

9.2.3 Effects of Symmetry-Breaking Weak Magnetic Field

We now proceed to the central theme, i.e., the magneto-resistance in these 3-d billiards. It should be noted that open ballistic 2-d billiards always share $g \propto \dfrac{e^2}{h}k_{\mathrm{F}}l$, independently of both integrability and lead orientations and **show no change in the exponent upon the application of a magnetic field** B. The negative magneto-resistance in 2-d billiards merely implies an increase of the coefficient. What types of novel phenomena will be expected in $SO(2)$-symmetric 3-d billiards under the perpendicular lead (A, C) orientation when a symmetry-breaking weak B field is applied

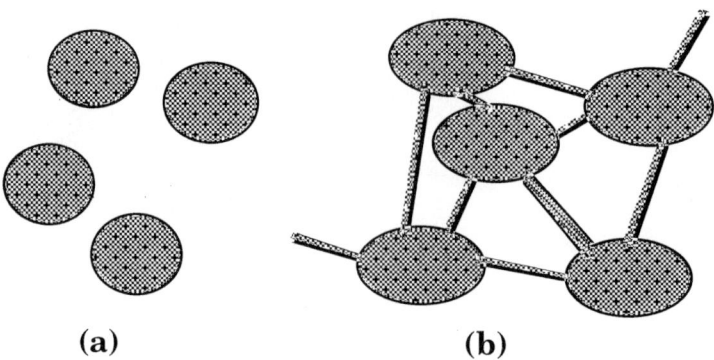

(a)   **(b)**

Fig. 9.7   Schematic phase-space structure. The dotted area represents chaotic regions: (a) isolated lakes of chaos (two degrees of freedom); (b) Arnold diffusion and Arnold web (three degrees of freedom).

along the X axis? In this case, the angular momentum is not the constant of motion, and electron motion is not planar.

In contrast to phase-space structures in 2-d systems where each of the chaotic zones generated by the weak B field is mutually separated by the Kolmogorov–Arnold–Moser (KAM) tori, those in 3-d systems consist of chaotic zones mutually connected by narrow chaotic channels that make up the **Arnold web (AW)** (see Fig. 9.7). In this case, incoming electrons in an arbitrary direction, which permeate (chaotic) zone by zone in an ergodic way through the AW, can show global and long-distant diffusion, i.e., **Arnold diffusion (AD)**. Therefore, the B field increases the transmission channels and the number of stationary points in the integration in (9.13), and the resultant integration in (9.15) over φ' and φ'' as well as θ' and θ'' indicates $g \propto \dfrac{e^2}{h}(k_{\mathrm{F}}l)^2$. In general, the width of the Arnold web (AW) depends on the magnitude of the B field: for the vanishing B field, no AW exists and each chaotic zone is isolated leading to 2-d like quantum transport; for the nonzero but sufficiently weak B field, the Arnold web (AW) is sufficiently thin to fully realize 3-d quantum transport. Therefore, we shall expect a more general result, $g \propto (k_{\mathrm{F}}l)^\beta$ with $1 \le \beta \le 2$, and the exponent β depends on the magnitude of the B field, namely, on the width of the AW. A crucial point is that the B field has increased the exponent in the power-law behavior of the semiclassical conductance in the case of the perpendicular lead orientation. This anomalous phenomenon is just the negative magneto-resistance beyond weak localization. Here, we employ the terminology of the **negative magneto-resistance** wherever the B field reduces the zero-field resistance.

On the other hand, for the parallel lead (Q, C) orientation, $g \propto \dfrac{e^2}{h}(k_{\mathrm{F}}l)^2$ is retained, since electron trajectories are not confined to a particular plane, irrespective

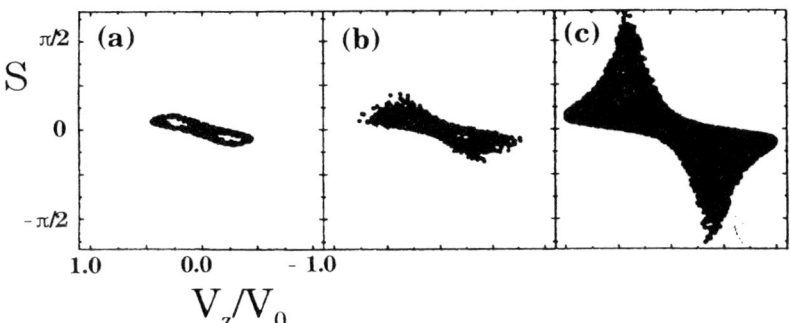

Fig. 9.8 Poincaré sections (X-Z plane) of the trajectories emanating from point A. The latitude S ($-\pi/2 \leq S \leq \pi/2$) and velocity component v_Z/v_0 (see the text for v_0) are chosen as Birkhoff coordinates. $S = 0$ and $S = \pi/2$ imply entrance A and exit C, respectively. $B = 0.02B_0$ and initial velocity $v_X/v_0 = 0.843, v_Y/v_0 = 0.538, v_Z/v_0 = 0$: (a)$P = 500$; (b)$P = 4100$; (c)$P = 16000$. P is the number of bounces at the outer billiard wall intersecting the X-Z plane. For the meaning of B_0, see the text. (Reproduced from[104].)

of the absence or presence of the B field; we can expect only the B field-induced increment of the proportionality constant, which implies the conventional negative magneto-resistance caused by suppression of the weak localization.

To numerically verify the above prediction, we shall choose a $SO(2)$-symmetric 3-d elliptic shell billiard system with the perpendicular lead orientation. This ellipsoid is a typical example of completely integrable $SO(2)$-symmetric 3-d billiards. The system in the presence of the symmetry-breaking field B parallel to the X axis will be examined. To understand the classical dynamics of electrons, we analyze, as a Poincaré surface of section, the X-Z plane with latitude S and velocity component $\dfrac{v_Z}{v^0}$ as Birkhoff coordinates. ($v^0 = \dfrac{\hbar\pi}{lm_e}$ is the velocity unit corresponding to the lowest mode.) For $B = 0$, the trajectory from the entrance A at $S \sim 0$ (the equator) with an arbitrary initial velocity vector lying outside the Z-Y plane proves to be confined to a torus, failing to reach the exit C at $S = \pi/2$ (north pole). Let η ($\equiv \Re/\Re_c$) be the ratio between the cyclotron radius $\Re_c$ and the characteristic length $\Re$ of the cross-sectional ellipse (see the caption of Fig. 9.5). Then the B field can be expressed as $B = \eta B_0$ with $B_0 = \dfrac{m_e v^0}{e\Re}$ (B_0 corresponds to $\Re$). In the case of a weak field ($B = 0.02B_0$), Poincaré sections of the trajectory with the initial velocity vector lying within the X-Y plane are given in Figs. 9.8(a)–(c) for increasing numbers of bounces at the outer billiard wall. One can find that as the B field is switched on, the orbit trajectory first leaves the initial torus and enters an outer layer, and repeating a similar process, it diffuses over more and more distant layers, exhibiting Arnold diffusion (AD). Even though the symmetry-breaking field is sufficiently weak to keep the orbit almost straight,

AD makes it possible for the trajectories with initial angular momenta $L_Z \neq 0$ to reach the exit C, i.e., $S = \pi/2$ (see Fig. 9.8(c)) and eventually to contribute to the conductance.

We proceed to the numerical calculation of the semiclassical transmission amplitude (9.13) for 3-d elliptic billiards with the perpendicular lead orientation. This is carried out without having recourse to the continuum-mode approximation, and a set of isolated trajectories s satisfying the rules below (9.14) are substituted into (9.13). In the present system, the 4×4 monodromy matrix $\mathbf{M}$ is defined as

$$\begin{pmatrix} \delta \mathbf{r}''_{\perp} \\ \delta \mathbf{p}''_{\perp} \end{pmatrix} = \mathbf{M} \begin{pmatrix} \delta \mathbf{r}'_{\perp} \\ \delta \mathbf{p}'_{\perp} \end{pmatrix}.$$

At the entrance we introduce a pair of mutually orthogonal coordinates $r'_{\perp 1}$ and $r'_{\perp 2}$ lying in the transverse plane perpendicular to the initial direction of each orbit. Similarly $r''_{\perp 1}$ and $r''_{\perp 2}$ are chosen at the exit. The first and last rotation angles in $\mathbf{M}$ should be determined by using $r'_{\perp 1}$, $r'_{\perp 2}$ and $r''_{\perp 1}$, $r''_{\perp 2}$, since, after folding the orbit, the initial and final coordinates should become identical. Using elements (M_{ij}) of matrix $\mathbf{M}$, we rewrite the factors necessary for the numerical computation of A_s and B_s in (9.13). For example,

$$\begin{aligned} \frac{\partial \theta'}{\partial x''} &= \frac{\partial(\theta', y'')}{\partial(r''_{\perp 1}, r''_{\perp 2})} \frac{\partial(r''_{\perp 1}, r''_{\perp 2})}{\partial(x'', y'')} \\ &= \frac{\cos \theta''}{p_{\mathrm{F}}} (\sec \varphi'' \csc \theta'' / M_{14} + \csc \varphi'' / M_{24}). \end{aligned}$$

Combining the factors as above with (9.13), the semiclassical conductance is calculated explicitly. The results, which include both the classical contribution and the quantum correction in an inseparable way, are plotted in Fig. 9.9. Although the data show small fluctuations, the fitted curves for their means show $g(B = 0.02B_0) - g(B = 0) \propto \frac{e^2}{h}(k_{\mathrm{F}}l)^2$ with $g(B = 0) \propto \frac{e^2}{h}k_{\mathrm{F}}l$, confirming the analytic result already suggested. The increase of the exponent in the k_{F} dependence of g upon switching on a weak field, which cannot be seen in 2-d cases, provides numerical evidence that the negative magneto-resistance in 3-d systems, i.e., the increment $g(B \neq 0) - g(B = 0)$ in Fig. 9.9, comes from AD in addition to a possible weak-localization correction.

We should note the following. (i) In the extremely weak field case when the width of the Arnold web (AW) is sufficiently small, we shall see $g \propto (k_{\mathrm{F}}l)^\beta$ with $1 \leq \beta < 2$. The supplementary numerical calculation suggests that, to have $\beta = 2$ in a very large k_{F} regime, $\Re/\Re_c$ should not be less than 10^{-2}. (ii) The role of a stopper is also essential. Reflection at an inner convex wall should increase the instability of orbits, which is favorable for the genesis of AD. (iii) To clearly observe AD, the symmetry-breaking field should destroy all constants of motion except for the energy. From this viewpoint, it is more advantageous to choose $SO(2)$-symmetric 3-d chaotic shell billiards generated from fully-chaotic 2-d billiards with a hollow by its rotation around the Z axis. The billiards of this kind, which are partially broken-ergodic, are experimentally more accessible than (completely broken-ergodic) elliptic shell billiard, and all conclusions for the latter should hold for the former as well.

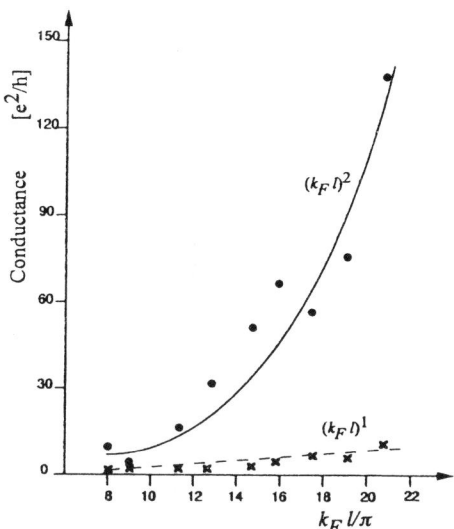

Fig. 9.9   Numerical results for semiclassical conductance versus Fermi wave number for the perpendicular (A, C) lead orientation. Filled circles and crosses indicate $g(B = 0.02B_0) - g(B = 0)$ and $g(B = 0)$, respectively, while solid quadratic and dashed linear lines, the corresponding mean curves (reproduced from[104]).

We have come to realize that the semiclassical conductance for three-dimensional (3-d) ballistic open billiards shows important features that cannot be seen in 2-d billiards. For partially or completely broken-ergodic 3-d billiards such as $SO(2)$ symmetric billiards, the dependence of the conductance g on the Fermi wave number k_F is dramatically changed by the lead orientation. The application of a symmetry-breaking weak magnetic field brings about mixed phase-space structures of 3-d billiards which ensures novel Arnold diffusion that cannot be seen in 2-d billiards. In the case of the perpendicular lead orientation, the Arnold diffusion increases the exponent in the k_F dependence of g in the phenomenon of the negative magneto-resistance.

9.3   Semiclassical Theory of Coulomb Blockade

The final section of this chapter is devoted to the semiclassical theory of Coulomb blockade. This phenomenon is originally attributed to a classical effect in quantum dots: the addition of an electron to an isolated dot (taken as a condenser with capacitance C) requires the charging energy $\dfrac{e^2}{2C}$. In nanostructures a sufficiently small capacitance enables single-electron transmission and leads to a large enough charging energy that prevents an extra electron from flowing in. A way to see Coulomb blockade in quantum transport is to bring in a **tunneling contact between the quantum dot and leads** and to measure conductance through the *almost closed* dot as a function of the gate voltage (or Fermi energy) that controls the electrostatic potential of the dot. As the point contacts are pinched off by using additional gates, the coupling becomes weaker and a barrier is effectively formed between the dot and the leads. In such *almost closed* dots, the charge is quantized. In contrast to the conductance without tunneling barrier that has so far been studied, the conductance is vanishing for most values of the gate voltage because the transmission of an electron from a lead to the dot is blocked by the Coulomb repulsion of electrons already existing in the dot. Let us call the difference of energies between the successive ground states with N and $N + 1$ electrons the **addition energy**. If, by tuning the gate voltage, the Fermi enegy happens to accord with the addition energy, the tunneling of an electron from one lead to the dot and cotunneling to another lead occurs. Therefore the conductance shows a spiky peak when the gate voltage has particular values that satisfy the above condition. Consequently, the conductance in the tunneling region consists of a series of conductance peaks as a function of the gate voltage or Fermi energy (see Fig. 9.10). This phenomenon is called the **Coulomb oscillation** – a feature very different from the continuous fluctuations of the conductance envisaged in Chapters 1–4. The spacing between successive peaks is roughly the charging energy $\dfrac{e^2}{2C}$, and the peak height is given by the conductance through the two tunnel barriers.

Precisely speaking, in small dots containing typically less than 20 electrons, the confining potential is harmonic-like, leading to regular dynamics of the electron and shell structure (Tarucha *et al.*[106]). This can be confirmed in the addition spectrum – the energy necessary to add an electron to the dot. Maxima in the addition spectrum are observed for the numbers of electrons corresponding to filled or half-filled valence harmonic-oscillator shells. These small numbers of electrons constitute an artificial atom or molecule, where the electron correlation plays an important role.

However, dots with $50 \sim 100$ electrons can be approximated by the mean-field model: they are described within the framework of single-particle dynamics in the mean-field but in the confining potential of the billiard system. For such dots the conductance and the addition spectrum show fluctuations when the shape of the dot or the magnetic field are changed. This is the statistical regime in which we are interested.

Chang *et al.*[108] measured the statistical distribution of peak heights in the conductance in the Coulomb blockade regime, where lifetime broadening is less than the mean level spacing and the electrostatic energy of the dot is larger than the applied

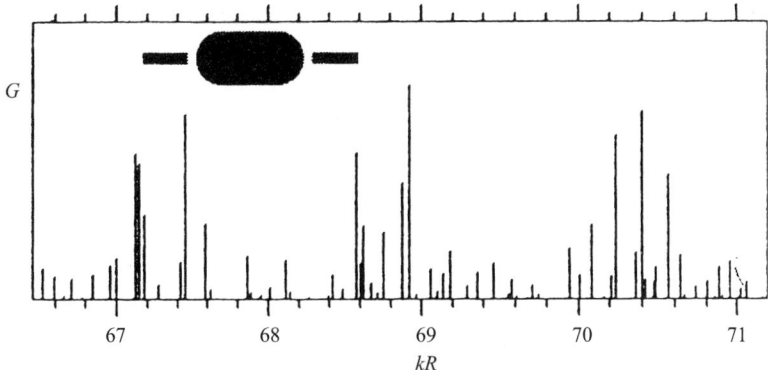

Fig. 9.10   The peak conductance from tunneling through subsequent energy levels for the stadium billiards in the inset. Each peak corresponds to its levels. In the abscissa, the wave number is scaled with use of R, the radius of the semicircle of the stadium (reproduced from[112]).

bias. They observed transport mediated by quantum tunneling through a single eigenstate of the dot. Since the Coulomb interaction in these region is comparable to the level spacing, the essential statistical quantity is the peak heights G_{peak} rather than the level spacings, while the peak positions are absolutely stable and reproducible. Chang *et al.* observed G_{peak} as a function of gate voltage and magnetic field by tuning the temperature and coupling to the leads. Then they proceeded to examine the statistical properties of G_{peak}. The non-Gaussian distribution they elucidated is the Porter–Thomas distribution with the square-root singularity near zero (see (9.29)). In a magnetic field greater than the correlation field, the breaking of time-reversal symmetry reduces the number of near-zero values of G_{peak}. Nevertheless, the distribution is still non-Gaussian. This contrasts with the Gaussian distribution of peak heights for typical metallic dots, where many levels are involved in the tunneling. Folk *et al.*[109], besides confirming the above result, showed that the autocorrelation function of peak height fluctuations agrees with the Lorentzian-squared form for the unitary ensemble. These results address several interesting issues to be explored further by theoreticians. Rigorous quantum-mechanical treatment of this phenomenon should result in: (i) fluctuations of the peak-to-peak separation due to the irregular one-particle energy spacing in quantum billiards with shapes in Fig. 1.4(a); (ii) peak height fluctuations due to a variation of coupling of the wavefunctions between dot and leads at the tunnel barrier.

Below, we shall concentrate on theme (ii). The experiment by Folk *et al.* indicates outstanding correlation of the intensities between the neighbouring peaks. It should be noted that while the random matrix theory certainly provides some universal feature of the statistics[111], it cannot convey the correlation among eigenfunctions, failing to give the nonuniversal feature of the peak-height fluctuations. Narimanov *et al.* argued that this correlation arises within an effective (mean-field) single-particle picture of the electrons in the quantum dot, and emphasized the role of short-time electron dynamics in describing the particular correlations between the conductance peak heights. Below we shall demonstrate how the semiclassical theory relates the quantum conductance peak height to the classical periodic orbits in the dot, following the scheme by Narimanov *et al.*[112].

9.3.1 Peak Height and Wavefunction

Consider a quantum dot placed close to two asymmetric leads 1 and 2, and let the partial decay widths Γ_1 and Γ_2 of a quantum level in the dot, which are due to the electron tunneling to leads 1 and 2, respectively. In the range, $\Delta E > k_B T > \Gamma_{1,2}$, the electrons can transmit through a single quantum level in the dot, and the net transmission rate Γ caused by double tunneling is given by

$$\Gamma = \frac{\Gamma_1 \Gamma_2}{\Gamma_1 + \Gamma_2}.$$

The Coulom blockade peak height is then expressed as (Beenekkar[105])

$$G_{\text{peak}} = \frac{e^2}{h} \frac{\pi}{2kT} \Gamma. \tag{9.16}$$

The partial width Γ_α ($\alpha = 1, 2$) is determined through the square of the tunneling matrix element $M^{\alpha \to d}$ between the lead (α) and the dot (d),

$$M^{\alpha \to d} = \frac{\hbar^2}{m_e} \int_S d\boldsymbol{r} \psi_\alpha(\boldsymbol{r}) \nabla_n \psi_d(\boldsymbol{r}), \tag{9.17}$$

where ψ_α and ψ_d are the lead ($\alpha = 1, 2$) and (mean-field) dot wavefunctions, respectively; ∇_n is the normal derivative at the tunnel barrier and the surface S denotes the edge of the quantum dot. From Fermi's golden rule, Γ_α is given by

$$\frac{1}{\hbar} \Gamma_\alpha[\psi_d] = \frac{2\pi}{\hbar} \sum_m \rho_{\alpha(m)} \left| M^{\alpha(m) \to d} \right|^2. \tag{9.18}$$

Here m represents each transverse subband in the lead α and $\rho_{\alpha(m)}$ is its density of states. The statistics of the conductance peak heights are available from the statistical properties of ψ_d, which will be mentioned below.

Let $P[\psi]$ be the distribution function which is needed. To determine the specific form of $P[\psi]$, we shall maximize Shannon's information entropy

$$H = -\int \mathcal{D}\psi P[\psi] \log P[\psi] \tag{9.19}$$

under a given constraint, where $P[\psi]\mathcal{D}\psi$ is the probability that a wavefunction $\psi(\mathbf{r})$ has a value between $\psi(\mathbf{r})$ and $\psi(\mathbf{r}) + d\psi(\mathbf{r})$ for any point inside the dot. In the present problem, the constraint imposed on the ensemble of wavefunctions is the two-point correlation function of an arbitrary wavefunction

$$C(\mathbf{r_1}, \mathbf{r_2}) = \int \mathcal{D}\psi P[\psi]\psi^*(\mathbf{r_1})\psi(\mathbf{r_2}). \tag{9.20}$$

The unique distribution $P[\psi]$ that maximizes the entropy (9.19) under the constraint (9.20) proves to take the Gaussian form

$$P[\psi] = A\exp\left[-\frac{\beta}{2}\int d\mathbf{r_1}\int d\mathbf{r_2}\psi^*(\mathbf{r_1})C^{-1}(\mathbf{r_1},\mathbf{r_2})\psi(\mathbf{r_2})\right]. \tag{9.21}$$

Here A is the normalization constant and C^{-1} is the functional inverse of the correlation function C; β stands for the system's symmetry ($\beta = 1$ and $\beta = 2$ correspond to systems with and without time-reversal invariance, respectively). In the semiclassical limit, the correlation function $C(\mathbf{r_2}, \mathbf{r_1})$ is determined by the semiclassical Green function G_{scl} as

$$C(\mathbf{r_2}, \mathbf{r_1}) = \frac{1}{\pi\overline{g}(E)}\mathrm{Im}G_{\mathrm{scl}}(\mathbf{r_2}, \mathbf{r_1}), \tag{9.22}$$

where $\overline{g}(E)$ is the mean part of the density of states in the dot. The semiclassical Green function itself was already exploited so far in this text, and its energy-average version is relevant here, which is expressed as

$$G_{\mathrm{scl}}(\mathbf{r_2}, \mathbf{r_1}) = G_0(\mathbf{r_2}, \mathbf{r_1}) + \frac{2\pi}{(2\pi i\hbar)^{3/2}}$$
$$\times \sum_{s(\mathbf{r_2},\mathbf{r_1})} \sqrt{D_s}\exp\left(\frac{i}{\hbar}S_s(\mathbf{r_2}, \mathbf{r_1}) - i\frac{\pi}{2}\mu_s\right)\exp\left(-\frac{\tau_s^2 W^2}{2\hbar^2}\right). \tag{9.23}$$

Here $s(\mathbf{r_2}, \mathbf{r_1})$ denotes each classical orbit from $\mathbf{r_1}$ to $\mathbf{r_2}$, and $S_s(\mathbf{r_2}, \mathbf{r_1})$ and τ_s are the corresponding action and period, respectively; the last exponential in (9.23) is due to the Gaussian averaging over an energy window of width W, and other indices and factors are the same as in the previous chapters.

9.3.2 Peak Height Distribution

Since the distribution for the wavefunctions in the dot is established in (9.21), the peak height distribution $P(G)$ is given by

$$P(G) = \int \mathcal{D}\psi_d P[\psi_d]\delta\left(G - G_{\mathrm{peak}}[\psi_d]\right), \tag{9.24}$$

where $G_{\mathrm{peak}}[\psi_d]$ is determined by (9.16)–(9.18). The peak height depends only on the wavefunction near the tunnel barrier at S as follows. Let us define

$$\psi(\mathbf{r}) = \begin{cases} \overline{\psi}(\mathbf{r}) & (\mathbf{r} \in S) \\ \hat{\psi}(\mathbf{r}) & (\mathbf{r} \notin S). \end{cases}$$

Then the conductance distribution is

$$P(G) = \int \mathcal{D}\overline{\psi} P_S[\overline{\psi}] \delta \left(G - G_{peak}[\overline{\psi}] \right). \tag{9.25}$$

The edge distribution P_S can be obtained from $P[\psi]$ in (9.21) by integrating out the value of $\hat{\psi}$:

$$P_S[\overline{\psi}] = A_S \exp\left[-\frac{\beta}{2} \int_S d\mathbf{q_1} \int_S d\mathbf{q_2} \overline{\psi}^*(\mathbf{q_1}) C^{-1}(\mathbf{q_1}, \mathbf{q_2}) \overline{\psi}(\mathbf{q_2}) \right]. \tag{9.26}$$

We assume $\overline{\psi}$ near the tunnel barrier to satisfy the Dirichlet boundary condition. Then in the narrow strip S with the coordinate y along the boundary of the dot and z in its normal direction, $\overline{\psi}$ becomes

$$\overline{\psi} = z\varphi(y) \equiv z \sum_{m=0}^{\infty} a_m \phi_m(y - y_l), \tag{9.27}$$

and the correlation function $C(\mathbf{q_2}, \mathbf{q_1})$ is

$$C(\mathbf{q_2}, \mathbf{q_1}) = z_2 z_1 \frac{1}{\pi \overline{g}(E)} \mathrm{Im} \partial_n G_{\mathrm{scl}}(y_2, y_1). \tag{9.28}$$

In (9.27), y_l is the the contact point and the dot wavefunction $\varphi(y)$ is expanded with use of the complete set of the wavefunctions corresponding to the transverse potential of the lead.

From (9.25)–(9.28), one obtains the semiclassical approximation to the conductance distribution of Porter–Thomas type,

$$P(G) = \left(\frac{2\pi}{\overline{G}} \right)^{1/2} \frac{1}{\sqrt{G}} \exp\left(-\frac{G}{2\overline{G}} \right). \tag{9.29}$$

Here $\overline{G}$ is the locally averaged conductance defined by

$$\frac{1}{\sqrt{\overline{G}}} = \frac{1}{\sqrt{\overline{G}_1}} + \frac{1}{\sqrt{\overline{G}_2}} \tag{9.30}$$

with $\overline{G}_\alpha$ $(\alpha = 1, 2)$ the locally averaged partial conductance calculated from (9.17) and (9.18) as

$$\frac{2hkT}{\pi e^2} \overline{G}_\alpha \left(= \overline{\Gamma}_\alpha \right) = \frac{2\pi \hbar^4}{m_e^2} \rho_{\alpha(0)} \int dy_1 \phi_{\alpha(0)}(y_1 - y_l) \phi_{\alpha(0)}^*(y_2 - y_l)$$

$$\times \frac{1}{\pi \overline{g}(E)} \mathrm{Im} \partial_n G_{\mathrm{scl}}(y_2, y_1). \tag{9.31}$$

Equation (9.31) includes the semiclassical Green function $G_{\mathrm{scl}}(y_2, y_1)$. Corresponding to the expression (9.23), $\overline{G}_\alpha$ can be decomposed into smooth and oscillating parts as

$$\overline{G}_\alpha = \overline{G}_\alpha^0 + \overline{\delta G}_\alpha.$$

Tunneling from the lead to the dot is dominated by the lowest transverse energy subband in the constriction between the lead and the dot, and the transverse potential in the tunneling region can be taken to be quadratic: $U_l \sim \omega_\perp (y - y_l)^2$. Then the transverse component of the lead wavefunction takes $\phi_{\alpha(0)} \sim \exp\left(-\dfrac{(y - y_l)^2}{2a_{eff}^2}\right)$ at the edge of the dot, and the effective width is $a_{eff} = \dfrac{\sqrt{\hbar}}{\sqrt[4]{2m_e\omega_\perp}} \sim \sqrt{\hbar}$. Such smallness of a_{eff} limits the integration in (9.31) to a semiclassically narrow region $y_2 \sim y_1 \sim y_l$, replacing the contribution of the open trajectories entering the Green function $G_{scl}(y_2, y_1)$ by their closed versions. Furthermore, in the resulting expression for $\overline{G}_\alpha$, the contribution of each closed orbit is accompanied with a damping factor exponentially small in Δp_y^2, where Δp_y is the momentum mismatch of the closed orbit, i.e., the change of transverse momentum after one traversal. The distribution of Δp_y is centered at zero with narrow width $\delta p_y \sim \dfrac{\hbar}{a_{eff}} \sim \sqrt{\hbar}$. After integration over Δp_y in (9.31) , only the contributions due to periodic orbits (PO's), i.e., orbits satisfying $y_2 \sim y_1 \sim y_l$ together with $\Delta p_y \sim 0$, can survive, as is the case of Gutzwiller's trace formula in Chapter 5. As a consequence, an explicit expression for the average conductance is

$$\overline{G} = \overline{G}_0 + \sum_{po} A_{po} \cos\left(\frac{S_{po}}{\hbar} + \phi_{po}\right), \tag{9.32}$$

where $(\overline{G}_0)^{-\frac{1}{2}} = (\overline{G}_1^0)^{-\frac{1}{2}} + (\overline{G}_2^0)^{-\frac{1}{2}}$ and A_{po} are also related to $\overline{G}_1^0$ and $\overline{G}_2^0$ as well as to the 2×2 stability matrix of PO. Thus the semiclassical conductance is characterized in a fashion similar to the semiclassical Kubo formula in Chapter 8, but **all the PO's here start at the edge of the dot close to the center of the lead.**

With use of $\delta G(E_m) \equiv G(E_m) - < G(E_n) >_n$ as a natural measure of the statistics of nearby peaks, the peak-to-peak correlation function is

$$\mathrm{Corr}_m[\delta G, \delta G] = \frac{< \delta G(E_{n+m})\delta G(E_n) >_n}{< [\delta G(E_n)]^2 >_n}$$

$$= \delta_{m,0} + (1 - \delta_{m,0})\frac{\sum_{po} A_{po}^2 \cos\left(\frac{\tau_{po}\Delta}{\hbar} m\right)}{4\overline{G}_0^2 + 3\sum_{po} A_{po}^2} \tag{9.33}$$

with Δ the interpeak spacing (Fermi energy $E_\mathrm{F} = m\Delta$). In the above, m is the peak number in case that energy is chosen as the tuning parameter that causes the peak height variation. One can expect a similar oscillatory behavior in the height of a given peak as a function of magnetic field. The semiclassical results in (9.29) and (9.33) are compared with the numerical quantal calculation in Fig. 9.11. The numerical correlation function (circles) agrees well with (9.33). The numerical probability distribution (histograms) is consistent with the semiclassical result (9.29).

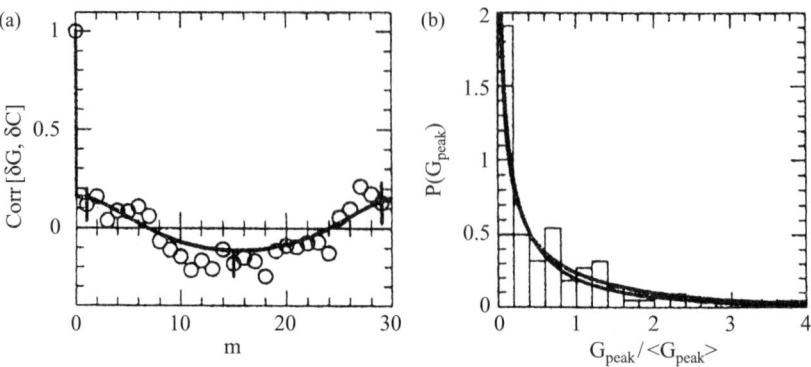

Fig. 9.11   Semiclassical and numerical conductance statistics corresponding to the stadium data in Fig. 9.10: (a) peak-to-peak correlation (circles for the numerics); (b) probability distribution of peak heights (histogram for the numerics) (reproduced from[112]).

Thus, the semiclassical theory of Coulomb blockade peak heights gives nonuniversal corrections to the predictions of random matrix theory, which are expressed in terms of the PO's starting from the contact point at the edge of the dot. Concerning the peak-height distribution in (9.29), it is locally a Porter–Thomas distribution but with the mean modulated by the periodic components. As for the peak-to-peak correlation function (9.33), in addition to the enhancement of the correlation of adjacent peaks, it oscillates as a function of peak number in a way characterized by PO's. Both formulas agree well with the experimental results[108–110].

Narimanov *et al.*'s theory was based on the idea that the peak width or the conductance peak due to a single resonance are proportional to the density of the eigenfunction inside the dot integrated over the contact region between the cavity and tunneling barriers. An alternative semiclassical description in resonance tunneling diodes was obtained by Bogomolny and Rouben[113] independently. Rather than evaluating the statistics of single-peak resonances, Budiyono and Nakamura[114] attempted to calculate the smoothed Coulomb blockade peak heights,

$$< G(E) > = \frac{e^2}{h} \frac{\pi}{2kT} \frac{1}{\Delta E} \int_{E-\Delta E/2}^{E+\Delta E/2} \Gamma(E')dE',$$

where the average is taken over a small but finite energy window ΔE containing as yet many resonances. They found that in the semiclassical limit the oscillating part of $< G(E) >$ can be expressed as a summation over periodic orbits (PO's) that bounce at the tunneling barriers. Because of the energy smoothing, the periods of PO's are less than the characteristic time $2\pi\hbar/\Delta E$. Therefore the result is nonuniversal and system-specific. Moreover, by calculating the local probability density inside the cavity, they derived Heller's wavefunction scars (i.e., enhanced probability density) along unstable periodic orbits that are well coupled to the entrance tunneling barrier[114]. These scars recur as system parameters are varied, whose frequency corresponds to the period of the PO's, i.e., one of the frequencies that determine the conductance peak height fluctuations.

In this chapter, we have treated three interesting themes bridging between nonlinear dynamics and quantum transport in mesoscopic billiards. Firstly, a triangular antidot lattice was investigated. We analyzed the semiclassical conductivity in a low-to-intermediate magnetic field. Within semiclassical periodic orbit theory, we found that the resistivity of the system yields a monotonic decrease with respect to the magnetic field. But, by including the effect of orbit bifurcations due to the overlapping between periodic orbits, several distinguished peaks of resistivity appear. The theoretical results have nicely explained both the locations and intensities of the anomalously large peaks observed in the experiment by the NEC group. Then, we proceeded to open three-dimensional (3-d) quantum dots. The mixed phase-space structure of 3-d billiards shows the Arnold diffusion that cannot be seen in 2-d billiards. A semiclassical conductance formula for ballistic 3-d billiards was derived. For partially or completely broken ergodic 3-d billiards such as $SO(2)$ symmetric billiards, the dependence of the conductance on the Fermi wave number is dramatically changed by the lead orientation. As a symmetry-breaking weak magnetic field is applied, the conductance shows a tendency to grow. In contrast to the 2-d case, the enhanced negative magneto-resistance near the zero field was attributed partly to the classical Arnold diffusion and partly to the quantum weak-localization correction. Finally, a semiclassical theory is developed on the statistical distribution for conductance-peak heights in the Coulomb blockade regime. In particular, the peak-to-peak correlation function was analyzed. Interestingly, the average conductance and the correlation function have reduced to the Gutzwiller-type formula expressed in terms of periodic orbits that start at the tunnel barrier locating at the edge of the quantum billiards.

Thus, while classical billiards launched by Kelvin are means by which to verify the foundations of statistical mechanics, quantum billiards (quantum dots and antidots) fabricated by nanotechnology provide means by which to capture quantal signatures of orbital bifurcations, Arnold diffusions, periodic orbits, and other interesting phenomena in nonlinear dynamics.

10

NONADIABATIC TRANSITIONS, ENERGY DIFFUSION AND GENERALIZED FRICTION

In this chapter, we shall investigate the energy diffusion in classically chaotic quantum systems. Until now the location of the billiard wall has been assumed fixed: a chaotic billiard system with a fixed boundary wall gives rise to a set of quantized energy levels. If the location of the wall is a tunable parameter, each energy level will change continuously as a function of the parameter (i.e., wall coordinates), leading to a parameter-dependent adiabatic energy spectrum. This spectrum is supposed to be described by parameter-dependent random matrix theory. Precisely speaking, the moving wall yields energy diffusion due to **nonadiabatic transitions** among energy levels, and consequently increases the energy of the quantum billiards which are initially at ground state. The resulting energy gain is nothing but the quantal origin of the frictional force acting on the moving wall, owing to the energy conservation of the whole system. The wall coordinates can be generalizable to macroscopic dynamical variables that couple with the quantum billiards, and the friction can also be generalized as well. In the Ohmic regime when the frictional force is proportional to the velocity of macroscopic variables, the proportionality coefficient implies a viscous constant, whose expression is shown to accord with the smooth part of semiclassical transport coefficients (e.g., Drude conductivity). The final section is devoted to discussions on future prospects of quantum chaos in nanotechnology.

10.1 What is Energy Diffusion?

In the previous chapters we have investigated time-independent stationary states. In this final chapter, we treat a time-dependent problem. Any quantum dot, to which a free-electron gas is confined, has so far been assumed to have fixed boundary wall with its position (say, X) being time-independent. If the wall were to move as illustrated in Fig. 10.1, what should happen to the free-electron gas confined by it? With X fixed, the solution of the time-independent Schrödinger equation affords a set of energy levels. A small change of X induces small shifts of individual energy levels. Repeating this procedure, one obtains a parameter(X)-dependent energy spectrum as in Fig. 10.2, which is called the **adiabatic spectrum**.

A more precise treatment would be as follows. When X is time-dependent, the electron gas in the billiard system should obey the **time-dependent Schrödinger equation** that causes nonadiabatic transitions among energy levels. As a consequence, the populations of individual energy states diffuse, leading to energy diffusion in the

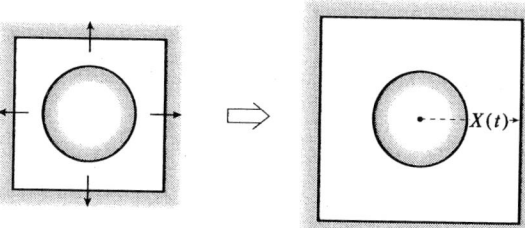

Fig. 10.1 Single Sinai billiard system with slowly-moving square walls. The circular (interior) wall is held fixed. A free-electron gas is confined between the circle and square walls. $X(t)$ denotes a macroscopic dynamical variable describing the distance between the origin and the square wall.

electron gas. One may generalize billiards to mesoscopic-scale quantum systems and similarly the wall coordinates to external macroscopic variables (e.g., magnetic or electric field) which control these quantum systems. In the following, we investigate energy diffusion phenomena in classically chaotic quantum systems. The energy conservation is premised for the whole system (quantum system + macroscopic dynamical variables). In other words, the time scale of energy relaxation to heat baths is assumed long enough to guarantee the quantum system to couple mechanically only with the macroscopic dynamical variables.

The behavior of energy diffusion in classically chaotic quantum systems is determined by two factors: (i) the level statistics of the adiabatic energy spectrum and (ii) the magnitude of the time derivative of external parameters. The energy diffusion inspired by nonadiabatic transitions increases the energy of a given system which is initially in the ground state. Noting energy conservation, this energy gain is compensated by the energy dissipation of the macroscopic dynamical variable. Thus, the energy diffusion in a microscopic quantum system is the quantum-mechanical origin of the frictional force on macroscopic dynamical variables (Hill and Wheeler[117]). The frictional force is also characterized by the above-mentioned factors (i) and (ii). Before proceeding to analyze the energy diffusion, we shall outline the level statistics of quantum systems.

10.2 What is Level Statistics?

Let us observe the adiabatic energy spectrum in Fig. 10.2 which looks like a complex spectrum of random matrix theory. Why does such a similarity arise? In classically integrable systems, the number of independent constants of motion agrees with the number of degrees of freedom. On quantization, there appears a set of quantum numbers corresponding to constants of motion. The energy manifolds with different

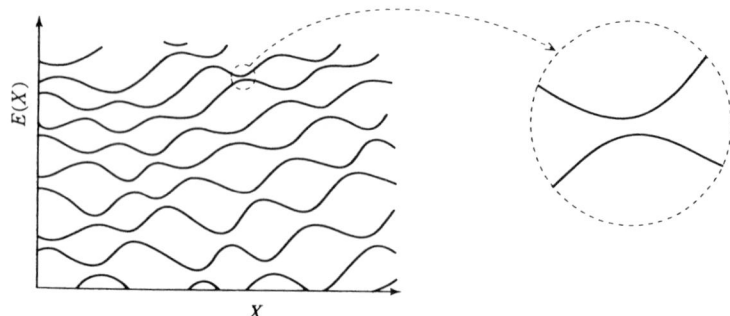

Fig. 10.2   Adiabatic energy spectrum of a classically chaotic system. The inset illustrates a single avoided level crossing.

quantum numbers are mutually intersecting, giving rise to a multitude of **energy level crossings**. The application of a nonintegrable perturbation on the integrable system will partially or completely destroy the constants of motion. In the corresponding quantum system, the mixing between manifolds with different quantum numbers alters all energy level crossings to **avoided level crossings**, generating **repulsion of nearby levels**. In consequence, a global feature of the energy spectrum seems equivalent to that of a complex spectrum of random matrix theory. There are several ways to verify this equivalence from the viewpoint of level dynamics[21,22] started by Dyson.

We here explain the framework of random matrix theory. Hermitian random matrix in $N \times N$ dimensions, $H\,(H^\dagger = H)$, has its matrix elements as

$$M \equiv N + \beta \frac{N(N-1)}{2}$$

independent stochastic variables. $\beta = 1$ for real symmetric matrices and $\beta = 2$ for complex Hermitian matrices. The former applies to the time-reversal symmetric system with no applied magnetic field, while the latter applies to the system with broken time-reversal symmetry.

How will the probability density $P(H)$ for the ensemble of matrices H be determined? Suppose G to be a set of transformations associated with the system's symmetry and $W\,(\in G)$ to be an arbitrary element. For any transformation within the ensemble

$$H' = WHW^{-1},$$

we impose the invariant probability $P(H')dH' = P(H)dH$. This invariance, noting the self-evident invariant measure $dH'(\equiv \prod_{i,j} dH'_{ij}) = dH(\equiv \prod_{i,j} dH_{ij})$, is nothing but the invariance of the probability density $P(H') = P(H)$. The last equality can only be satisfied by employing a Gaussian form

$$P(H) = c \exp\left(-a \operatorname{Tr} H^2\right).$$

The real symmetric ensemble H invariant under the orthogonal transformation G is called a Gaussian orthogonal ensemble (GOE), and the complex Hermitian ensemble H invariant under the unitary transformation G is called a Gaussian unitary ensemble (GUE). Similarly there is the Gaussian symplectic ensemble (GSE) in which the Hamiltonian H with Kramers' degeneracy commutes with the time-reversal operator K but with $K^2 = -1$.

$P(H)$ has M stochastic variables, but a suitable transformation to diagonalize H can reduce $P(H)$ to the joint probability density that has N eigenvalues as new stochastic variables:

$$P(\{E_n\}) = \text{const.} \times \exp\left(-\frac{E_1^2 + E_2^2 + \cdots + E_N^2}{4a^2}\right) \prod_{n>m} |E_n - E_m|^\beta.$$
$$(10.1)$$

Equation (10.1) can further be reduced to a **nearest-neighbor level spacing distribution**. Consider the simplest case $N = 2$ with two eigenvalues $E_1, E_2 \ (> E_1)$. Then, replace E_1, E_2 by their sum

$$\overline{E} = E_2 + E_1$$

and difference

$$S = E_2 - E_1,$$

and finally integrate the $N = 2$ version of (10.1) over $\overline{E}$. We then obtain the Wigner distributions

$$P_{\text{GOE}}(S) = \frac{\pi}{2} S \exp\left(-\frac{\pi}{4} S^2\right), \tag{10.2a}$$

$$P_{\text{GUE}}(S) = \frac{32}{\pi^2} S^2 \exp\left(-\frac{4}{\pi} S^2\right) \tag{10.2b}$$

for ensembles GOE and GUE, respectively, with $S(0 < S < \infty)$ the level spacing. Equations (10.2a) and (10.2b) show a peak at a nonzero S value, reflecting level repulsion.

On the other hand, quantized integrable systems like circle and square billiards show no level repulsion, and level spacings S for the desymmetrized spectrum obey the **Poisson distribution**

$$P(S) = \exp(-S). \tag{10.3}$$

Further, almost all nonintegrable systems have phase space which is occupied by both chaos and KAM tori, and the corresponding quantum systems have **intermediate level spacing distributions** interpolating between Wigner and Poisson distributions. However, there seems to be no universal distribution in this intermediate region. Therefore, in our investigation of the energy diffusion, we shall concentrate on the fully chaotic systems whose quantum counterparts show the Wigner level-spacing distribution in a wide parameter (X) range. The description below follows Wilkinson's scheme[119, 120].

10.3   Energy Diffusion: Landau–Zener regime

A classically chaotic quantal system should show energy diffusion, when being coupled with some external macroscopic **dynamical variable** $X(t)$ (see Fig. 10.3). The feature of such energy diffusion depends on the rate of temporal change of $X(t)$, and will be classified into two typical regimes, i.e., **Landau–Zener** and **linear response**[119, 120].

We first investigate the Landau–Zener regime. In this regime, the rate of temporal change of $X(t)$ is so small that the energy diffusion is dominated by Landau–Zener transitions between adjacent states. Each avoided level crossing is characterized with energy gap ΔE and (relative) level gradient A (i.e., the difference between gradients of a pair of nearby levels, see the inset in Fig. 10.3). In the vicinity of an avoided crossing, the level spacing

$$\varepsilon(X) = E_{n+1}(X) - E_n(X)$$

is given by

$$\varepsilon(X) = (\Delta E^2 + A^2 X^2)^{1/2}. \tag{10.4}$$

Using Landau–Zener formula, the transition rate between the nearby levels is expressed as

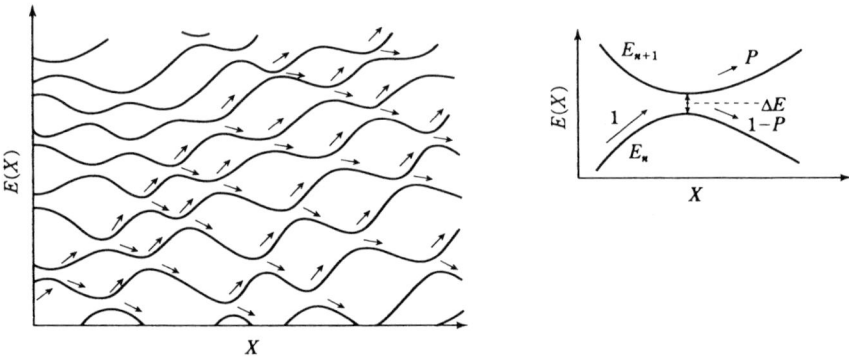

Fig. 10.3   Energy diffusion induced by nonadiabatic transition. The inset indicates Landau–Zener transitions between adjacent states.

$$P = \exp\left(-\frac{\pi \Delta E^2}{2A\hbar |\dot{X}|}\right) \tag{10.5}$$

with $\dot{X} \equiv \dfrac{dX}{dt}$. (10.5) implies the transition to be suppressed ($P \sim 0$) for a small enough $|\dot{X}|$ and to become notable ($P \sim 1$) for increasing $|\dot{X}|$. While time elapses, X traces a multitude of avoided crossings where Landau–Zener transitions occur. We also introduce $\mathcal{P}(A, \Delta E)$ as a distribution for avoided crossings with ΔE and A lying in the unit parameter (X) interval. The mean value of the transition rate per unit time is evaluated as

$$R = \dot{X} \int_0^\infty dA \int_0^\infty d\Delta E \mathcal{P}(A, \Delta E) \exp\left(-\frac{\pi \Delta E^2}{2A\hbar |\dot{X}|}\right), \tag{10.6}$$

where $\dot{X}$ denotes the area that X traces during unit time. Provided the random matrix theory (GOE, GUE) is applicable, the distribution for avoided crossings is written explicitly.

Assume two levels $E_{n+1}(X)$ and $E_n(X)$ become close or almost degenerate at $X = X_0$. The neighborhood of $X = X_0$ is then described by an effective Hamiltonian,

$$H = \begin{pmatrix} e^+ & h \\ h^* & e^- \end{pmatrix}. \tag{10.7}$$

In (10.7) the diagonal element is

$$e^- = E_n(X_0) + \left(\frac{\partial H}{\partial X}\right)_{n,n}(X - X_0), \tag{10.8}$$

$$e^+ = E_{n+1}(X_0) + \left(\frac{\partial H}{\partial X}\right)_{n+1,n+1}(X - X_0), \tag{10.9}$$

and the nondiagonal one is

$$h = \left(\frac{\partial H}{\partial X}\right)_{n,n+1}(X - X_0) = h_1(X - X_0). \tag{10.10}$$

With use of the prescription

$$e_0 = E_{n+1}(X_0) - E_n(X_0), \tag{10.11}$$

$$e_1 = \left(\frac{\partial H}{\partial X}\right)_{n+1,n+1} - \left(\frac{\partial H}{\partial X}\right)_{n,n}, \tag{10.12}$$

and the level spacing at an arbitrary X, which is a generalization of (10.4), is given from (10.7) as

$$\varepsilon(X) = \left\{\Delta E^2 + A^2(X - X_0 - \lambda)^2\right\}^{1/2} \tag{10.13}$$

where A, λ, and ΔE are simple functions of e_0, e_1, and h_1 (λ is a shift of the avoided crossing point from the origin of expansions in (10.8)–(10.10)). e_0, e_1, and h_1 themselves are stochastic variables obeying the distribution of random matrices. In particular, e_0 obeys the level-spacing distribution in (10.2a), (10.2b), and specifically in the case of GOE,

$$P(e_0) = \frac{\pi}{2} e_0 \exp\left(-\frac{\pi}{4} e_0^2\right). \tag{10.14}$$

Similarly, $e_1(h_1)$ is assumed to satisfy the Gaussian distribution with variance $4\sigma^2(\sigma^2)$, as is the case of diagonal (nondiagonal) elements in random matrices.

In terms of distributions for e_0, e_1, h_1, one can define the distribution for avoided crossing with $A, \Delta E$ as

$$\mathcal{P}(A', \Delta E')dA'd\Delta E'$$
$$= \lim_{\Delta X \to 0} \frac{1}{\Delta X} \int de_0 \int de_1 \int dh_1 P(e_0)P(e_1)P(h_1)\delta(A' - A)$$
$$\times \delta(\Delta E' - \Delta E)\Theta(\lambda)\Theta(\Delta X - \lambda)dA'd\Delta E'. \tag{10.15}$$

Here $\Theta(\cdot)$ is the Heaviside step function. After integrations, (10.15) reduces to

$$\mathcal{P}(A, \Delta E) \sim \Delta E^{\beta-1} A^{\beta+1} \exp\left(-\frac{A^2}{8\sigma^2}\right). \tag{10.16}$$

(β is 1 and 2 for GOE and GUE, respectively.) Substitution of (10.16) into (10.6) yields the expectation for the transition rate,

$$R \sim \hbar^{\beta/2} g_0^{\beta+1} \sigma^{\beta/2+1} |\dot{X}|^{\beta/2+1}. \tag{10.17}$$

Here g_0 is the density of states.

How will the occupation probability for each adiabatic state evolve under the time-dependent parameter $X(t)$? Let f_n be the occupation probability for the n-th adiabatic state. In the circumstances that only the transition between nearest-neighboring levels is permissible and that the transition rate is of Landau–Zener type, f_n is governed by the master equation

$$\frac{df_n}{dt} = (-2f_n + f_{n+1} + f_{n-1})R, \tag{10.18}$$

which, in the continuum limit, reduces to the **diffusion equation in energy space**,

$$\frac{\partial f}{\partial t} = \frac{R}{g_0^2} \frac{\partial^2 f}{\partial E^2} \equiv D\frac{\partial^2 f}{\partial E^2}. \tag{10.19}$$

From the second equality together with (10.17), one finds the diffusion constant D to be expressed as

$$D_{\text{GOE}} = \frac{R}{g_0^2} \sim \hbar^{1/2}\sigma^{3/2}|\dot{X}|^{3/2}, \tag{10.20a}$$

$$D_{\mathrm{GUE}} \sim \hbar g_0 \sigma^2 \dot{X}^2 \tag{10.20b}$$

for GOE and GUE, respectively. The point to be emphasized is that the exponent in the velocity-dependence of the diffusion constant varies depending on the universality class of the level-spacing distribution. The assertion in this section is valid so long as the magnitude of the velocity ($|\dot{X}|$) is sufficiently slow to guarantee the Landau–Zener transition rate $P \ll 1$ in (10.5), i.e., for

$$\hbar g_0^2 \sigma |\dot{X}| \left(\sim \frac{\hbar A |\dot{X}|}{\Delta E^2} \right) \ll 1. \tag{10.21}$$

10.4 Energy Diffusion: Linear-Response Regime

We shall proceed to the case opposite to that of the previous section. When $|\dot{X}|$ becomes larger so that transitions to all distant levels make a significant contribution to the energy diffusion, the analysis based on the **Kubo–Greenwood method** plays a vital role. The eigenvalue equation at each instant of time,

$$H(X(t))|n(X(t))\rangle = E_n |n(X(t))\rangle, \tag{10.22}$$

provides the so-called adiabatic eigenstates $|n(X(t))\rangle$ and eigenvalues $E_n(X(t))$, in terms of which the time-dependent state function $|\Psi(t)\rangle$ can be expanded as

$$|\Psi(t)\rangle = \sum_n c_n(t) \exp(i\varphi_n(t))|n(X(t))\rangle, \tag{10.23a}$$

$$\varphi_n(t) = \frac{1}{\hbar} \int^t dt' E_n(X(t')). \tag{10.23b}$$

Using (10.23a) and (10.23b) in the **time-dependent Schrödinger equation**, one obtains the equation for the coefficients c_n,

$$\dot{c}_n = \dot{X} \sum_{m(\neq n)} \frac{\left(\frac{\partial H}{\partial X}\right)_{nm}}{E_n - E_m} \exp\big(i(\varphi_n(t) - \varphi_m(t))\big) c_m. \tag{10.24}$$

For an increment Δc_n after a small time Δt, the variation of the occupation probability $f_n(\equiv |c_n(t)|^2)$ for the n-th adiabatic state is

$$\begin{aligned} \Delta f_n &= |c_n + \Delta c_n|^2 - |c_n|^2 \\ &= c_n^* \Delta c_n + c_n \Delta c_n^* + |\Delta c_n|^2. \end{aligned} \tag{10.25}$$

We here employ the random-phase approximation (RPA) which ignores the first two terms on the right-hand side of the last equality by assuming a rapid phase change on a time scale much less than Δt. Taking the average of (10.25) over initial values and applying the RPA again to suppress nondiagonal terms in $\langle |\Delta c_n|^2 \rangle$, we have

$$\langle \Delta f_n \rangle = \langle |\Delta c_n|^2 \rangle = \sum_{m(\neq n)} \frac{\dot{X}^2}{(E_n - E_m)^2} \Delta t \, \langle |c_m(0)|^2 \rangle$$

$$\times \int_{-\infty}^{\infty} d\tau \exp \left(\frac{i}{\hbar} (E_n - E_m)\tau \right) C(\tau), \qquad (10.26)$$

where $C(\tau)$ means the time-correlation function defined by

$$C(\tau) \cong \frac{\left\langle \left(\frac{\partial H}{\partial X} \right)_{nm} (0) \left(\frac{\partial H}{\partial X} \right)_{nm} (\tau) c_m(0) c_m(\tau) \right\rangle}{\langle |c_m(0)|^2 \rangle}. \qquad (10.27)$$

The above average is taken over n and m. In the limit $\Delta t \ll 1$, (10.26) proves to be reduced to a master equation for the occupation probability f_n,

$$\frac{df_n}{dt} = \sum_{m(\neq n)} R_{nm}(f_m - f_n). \qquad (10.28)$$

Here $R_{nm}(= R_{mn})$ is the **effective transition rate** given by

$$R_{nm} = \frac{\dot{X}^2}{(E_n - E_m)^2} \int_{-\infty}^{\infty} d\tau \exp \left(\frac{i}{\hbar} (E_n - E_m)\tau \right) C(\tau). \qquad (10.29)$$

This expression can be understood by comparing (10.26) and (10.28). Following the procedure in the previous section, let us regard n as a continuum variable and take the continuum limit of (10.28), and then we obtain the energy diffusion equation

$$\frac{\partial f(n)}{\partial t} = \int_{-\infty}^{\infty} dn' R(n') \left[f(n + n') - f(n) \right]$$

$$\equiv D \frac{\partial^2 f(n)}{\partial n^2}. \qquad (10.30)$$

With use of (10.29), the diffusion constant in this case reads as

$$D = \frac{1}{2} \int_{-\infty}^{\infty} dn' R(n') n'^2$$

$$= \frac{1}{2} g_0 \dot{X}^2 \int_{-\infty}^{\infty} d\tau C(\tau) \int_{-\infty}^{\infty} d\Delta E \exp \left(\frac{i}{\hbar} \Delta E \tau \right)$$

$$= \pi \hbar g_0 \dot{X}^2 \int_{-\infty}^{\infty} d\tau C(\tau) \delta(\tau)$$

$$= \pi \hbar g_0 \sigma^2 \dot{X}^2. \qquad (10.31)$$

To reach the last equation, $C(0) = \sigma^2$ was employed since $C(0)$ is the variance of nondiagonal elements $\left(\frac{\partial H}{\partial X} \right)_{n,m}$. From (10.31) we find that the result in the linear-response regime is that irrespective of the universality class of level-spacing distributions, the exponent in the velocity dependence of the diffusion constant is always two.

This conclusion holds good in the case that the transitions to all distant levels are permissible and that the double integral in (10.31) has a non-vanishing value, namely, when $|\dot{X}|$ satisfies

$$\hbar g_0^2 \sigma |\dot{X}| \gg 1 \quad \text{and} \quad \hbar g_0 \sigma |\dot{X}|/\Delta E_c \ll 1. \tag{10.32}$$

The former inequality is just opposite to (10.21), while the latter one implies that the decay time $(\tau_c \sim (g_0 \sigma |\dot{X}|)^{-1})$ of $C(\tau)$ in (10.27) should be much larger than the Thouless time $(\hbar/\Delta E_c)$. In this and the previous sections, we have argued that the energy diffusion inspired by nonadiabatic transitions in classically chaotic quantum systems demonstrates very rich features depending on the level statistics and the rate of temporal change of the external macroscopic variables X coupled to the quantum systems.

10.5 Frictional Force due to Nonadiabatic Transition

The next question to be raised is from where should the energy increment due to the energy diffusion be supplied. We are assuming that the time scale of energy relaxation to heat baths is long enough to guarantee the quantum system to mechanically couple only with the macroscopic dynamical variables. Under these circumstances, it is natural to consider that an external dynamical variable should receive the frictional force F_d from the quantal system and that the work done by this frictional force should be used to induce the energy diffusion under investigation. As everybody knows, the frictional force of classical origin is proportional to the velocity $\dot{X}$ of a dynamical variable. On the contrary, the present frictional force is of quantal origin. Is its dependence on the velocity different from the classical counterpart?

Consider a noninteracting Fermi gas and let $f(E, t)$ be the time(t)-dependent occupation probability for each single-particle energy level E. The initial distribution $f(E, t = -\infty)$ is supposed to be of Fermi–Dirac type. At low enough temperature, the total energy for the Fermi gas is written as

$$\overline{E}(t) \approx g_0(E_F) \int dE E f(E, t), \tag{10.33}$$

where $g_0(E_F)$ is the density of states at Fermi energy E_F. As X changes with time, $\overline{E}(t)$ increases with rate

$$\begin{aligned} \frac{\partial \overline{E}}{\partial t} &= g_0(E_F) \int dE E \frac{\partial f}{\partial t} \\ &= g_0(E_F) D \int dE E \frac{\partial^2 f}{\partial E^2} \\ &= g_0(E_F) D, \end{aligned} \tag{10.34}$$

where a partial integration was carried out to obtain the final expression. Since the negative of the work done by the frictional force F_d should be equal to the energy gain in the Fermi gas, F_d is defined by

$$F_d = \frac{\partial \overline{E}}{\partial X} = \frac{\partial \overline{E}}{\partial t} \frac{1}{\dot{X}} = \frac{g_0 D}{\dot{X}}. \tag{10.35}$$

Substitution of (10.20a), (10.20b), and (10.31) into (10.35) results in the important conclusion: in the Landau–Zener regime (slow-velocity regime),

$$F_d \propto \dot{X}$$

for GUE, and

$$F_d \propto \dot{X}^{1/2}$$

for GOE, indicating the frictinal force to depend on the universality class of the level-spacing distribution. In the linear-response regime (fast-velocity regime), on the other hand, $F_d \propto \dot{X}$ irrespective of the level statistics, which is the same formula as that for the standard frictional (viscous) force envisaged in Ohm's law in electrical resistance.

Concerning the linear-response regime, the frictional force is available more directly without having recourse to diffusion coefficients, as shown below. Define the expectation value of the force F as

$$F = \mathrm{Tr}\left(\hat{\rho}(t)\frac{\partial \hat{H}}{\partial X}\right). \tag{10.36}$$

The density operator for a quantal system $\hat{\rho}(t)$, when expanded in terms of adiabatic eigenstates, is a sum of the diagonal part and a correction proportional to $\dot{X}$:

$$\hat{\rho}(t) = \sum_n |n(t)\rangle f_n \langle n(t)| + \hat{\rho}_1 \dot{X}. \tag{10.37}$$

Using (10.37) in (10.36), we find

$$F = F_0 + F_d \equiv \sum_n f_n \frac{\partial E_n}{\partial X}\bigg|_{X(t)} + \mu \dot{X}. \tag{10.38}$$

The above equality means that F is a sum of the (reversible) adiabatic force F_0 and and an (irreversible) nonadiabatic correction F_d. μ stands for the viscosity coefficient, and can be reduced to the electrical conductivity in the special case of quantum transport, as explained below. The form for F_d is nothing but the one already obtained via diffusion constants in the linear-response regime.

Combining (10.35) with (10.31) and assuming that the occupation probability f drops sharply at the Fermi energy E_F, one finds

$$\mu = \pi \hbar g_0^2 \sigma^2(E_F). \tag{10.39}$$

As noted below (10.14), $\sigma^2(E_F)$ means the variance of matrix elements for $\dfrac{\partial \hat{H}}{\partial X}$. To define more explicitly,

$$\sigma^2(E_F) = \frac{1}{g_0^2} \sum_{n,m} \left|\left(\frac{\partial \hat{H}}{\partial X}\right)_{nm}\right|^2 \delta(E_F - \frac{1}{2}(E_n + E_m))\delta(E_n - E_m).$$

$$(10.40)$$

If the quantum system characterized by random matrices has the underlying classical chaotic dynamics, $\sigma^2(E_F)$ can be expressed in terms of the time correlation,

$$C_A(E_F, t) = \int d\boldsymbol{p}\, d\boldsymbol{q}\, A(\boldsymbol{p}, \boldsymbol{q}) A(\boldsymbol{p}(t), \boldsymbol{q}(t)) \delta(E - H(\boldsymbol{p}, \boldsymbol{q})).$$

$$(10.41)$$

Here, for the system with d degrees of freedom, the above integration is made over the phase space in 2-d dimensions. $A(\boldsymbol{p}(t), \boldsymbol{q}(t))$, corresponding to $\dfrac{\partial \hat{H}}{\partial X}$, stands for a classical observable conjugate to X. In this semiclassical regime, μ in (10.39) is rewritten as[119, 120]

$$\mu = \frac{g_0}{\Omega} \int_{-\infty}^{\infty} dt\, C_A(E_F, t).$$

$$(10.42)$$

Ω means the volume of the energy shell with energy E_F, and is related to the density of states as $g_0 = \dfrac{\Omega}{(2\pi\hbar)^d}$. As is the case of the Gutzwiller trace formula, the formulas in (10.41), (10.42) tell us that information on nonlinear classical dynamics can determine the viscosity coefficient of the nonadiabatic-transition-induced frictional force.

Equations (10.41) and (10.42) also give a general semiclassical formula for the quantum transport coefficient. In fact, the Drude formula (see Chapters 7 and 8) for conductivity can be interpreted as follows. Consider a conducting Aharonov–Bohm ring with a time-dependent magnetic flux, $\Phi(t)$, penetrating through its hollow. Once $\Phi(t)$ is taken as an external macroscopic variable X, its velocity $\dot{X}$ and force $\dfrac{\partial \hat{H}}{\partial X}$ correspond to the electric field $E\,(= \dot{\Phi})$ and current $J\,\left(= \mathrm{Tr}\left(\hat{\rho}\dfrac{\partial \hat{H}}{\partial X}\right)\right)$, respectively.

In the adiabatic limit, the single-electron energy spectrum varies periodically (with period of the flux quanta $\Phi_0\,\left(= \dfrac{ch}{e}\right)$) as a function of the flux. If the ring were replaced by fully chaotic Sinai billiards (see Chapter 7), the adiabatic quantal energy spectrum will obey the predictions of random matrix theory: the level statistics here is of GUE type except for the case that the magnetic flux takes integer or half-integer multiples of Φ_0. Application of the general result just below (10.35) leads to Ohm's law

$$J = \sigma E \qquad\qquad (10.43)$$

in each of the Landau–Zener and linear-response regimes, where σ stands for the electrical conductivity. In particular, in the linear-response regime, the formula in (10.42) and (10.43) is applicable and one has

$$\sigma = \frac{g_0}{\Omega} \int_{-\infty}^{\infty} dt C_J(E_F, t). \tag{10.44}$$

Ω is the volume of the energy shell and $C_J(E, t)$ is the time correlation function of the current $J(\boldsymbol{p}(t), \boldsymbol{q}(t))$, namely,

$$C_J(E_F, t) = \int d\boldsymbol{p} d\boldsymbol{q} J(\boldsymbol{p}, \boldsymbol{q}) J(\boldsymbol{p}(t), \boldsymbol{q}(t)) \delta(E_F - H(\boldsymbol{p}, \boldsymbol{q})). \tag{10.45}$$

Equations (10.44) and (10.45) are just the Drude formula: the smooth part of the semiclassical conductivity (see Chapter 8) is now reproduced from the viewpoint of a general framework of nonadiabatic transition and energy diffusion. One may also reproduce the quantum correction by treating the deterministic Hamiltonian which has the classical chaotic limit.

10.6 Future Prospects

As addressed at the beginning of this book, the physics of billiards was started with Kelvin's monumental lecture one century ago, and it has been innovated by progress in nanotechnology since the 1990s. Billiards are nowadays fabricated as quantum dots and antidots. Thanks to the semiclassical theory, we have come to understand quantum fluctuations in quantum transport on the basis of the classical picture. The knowledge of nonlinear dynamics, namely, transmission and reflection orbits, orbit bifurcations, and Arnold diffusion as well as periodic orbits, plays an important role in our understanding of quantum transport in quantum dots and antidots. The experimental and theoretical themes in this field, which concentrate on the investigation of the *quantal manifestation of chaos*, constitute a paradigm of *quantum chaos*.

There are other promising subjects worthy of intensive investigations. Among them, the effect of diffraction is important, and this goes beyond the framework of realistic classical orbits. *Diffractive orbits* will play an interesting role in the case that a billiard system has angular corners and/or (touchable) Aharonov–Bohm flux (Sieber[121], Bogomolny *et al.*[122]): the diffractive correction can be of the order of the isolated-orbit contributions. An approximation based on Kirchhoff's theory is also proposed by Bogomolny *et al.* At present, however, the activities are quite limited to the mathematical aspect with little attention to quantum transport in quantum dots. It is highly desirable to apply the new method and result to the analyses of conductance and conductivity.

The semiclassical analysis of the renormalization-group approach for the transition from the last KAM tori to global chaos would also be a challenging theme, since the corresponding studies on the classical level are well accumulated. In this context, the effect of orbit bifurcations on the semiclassical propagator plays a crucial role. As described in Chapter 7, the experiments on single quantum dots with a weak magnetic field convey the emergence of self-similar magneto-conductance fluctuations, which indicates a cascade of pitchfork bifurcations of periodic orbits in the underlying classical dynamics. In fact, in some range of Fermi energy just below a threshold value, one can expect a scaling regime where the classical renormalization-group approach is applicable.

On the other hand, a lot of experiments are devoted to artificial atoms or molecules (Tarucha *et al.*[106], Kouwenhoven *et al.*[107]) which confine a few through several tens of electrons into a quantum dot, as explained in Section 9.3. Here Coulomb interaction among electrons is essential (i.e., the mean-field theory breaks down) and the number of degrees of freedom for this system is greater than or equal to 3. Let us consider the case that the confinement is perfect with neither leaky channels nor quantum tunneling through leads. Then, even a circular quantum dot containing correlated electrons has the degrees of freedom larger than the number of independent constants of motion. The system can be classically chaotic in more-or-less high energy states, whose quantum counterpart is a manifestation of quantum chaos. The observation of quantal high energy states via optical absorption would be of interest.

This textbook suppressed the dynamics of the spin degrees of freedom that is also becoming important in the field of *spin-electronics* or *spintronics*. The semiclassical theory of spintronics in quantum dots has not yet been established, and would be a promising subject in future. There is a growing number of other interesting subjects emanating from chaos in quantum dots, e.g., quantum ratchets working in nonequilibrium, nonhyperbolicity-induced dynamical tunneling, lead-assisted wavefunction scars, effect of chaos on quantum computers, to mention a few. See the many articles in the Proceedings of Nobel Symposia[30] and also the collection of review articles on the recent experiments[42].

Systems treated so far, except in this chapter, are all in (time-independent) stationary states. Consider the quantum dynamics of an electron within a closed stadium billiard system. Its wave packet quickly diffuses during a short transient period, and eventually shows regular motion, where it is difficult to see the classical–quantal correspondence. Likewise, quantum dynamics for a system with many degrees of freedom shows no interesting behavior corresponding to Arnold diffusion, Levy flight, etc. Unless quantum coherence is kept suppressed by continuous coupling with an external environment, the quantum dynamics has no possibility to exhibit symptoms of chaos except during an initial short time. These facts are natural consequences of the linearity of the time-dependent Schrödinger equation, but are quite mysterious issues, considering Bohr's correspondence principle. Below, we shall present some speculation on this cloud hanging over quantum mechanics, without being too serious for the grounds of an argument.

In the fourth article in his series of papers to establish wave-mechanics, Schrödinger invented the time-dependent Schrödinger equation, closely following the framework of linear wave-optics. This equation is motivated to describe de Broglie's matter–wave corresponding to a particle undergoing periodic or quasi-periodic motion. On the contrary, a particle undergoing chaotic motion has no such regularity, and therefore a nonlinear wave-equation such as the nonlinear Schrödinger equation might be a more suitable counterpart. In fact Weinberg[123,124] previously proposed a nonlinear variant of quantum mechanics to give a consistent explanation for the irregular precession of a nuclear spin of Penning-trapped Beryllium ions in the presence of a magnetic field. His colleague, Polchinski[125], pointed out that Weinberg's nonlinear quantum mechanics should yield a conversion from virtual to real breakdown of locality in the Einstein–Podolsky–Rosen *gedanken* experiment and that the velocity of signal transmission can be infinite, beyond the velocity of light. However, this kind of theoretical activity ceased afterwards, since comprehensive experiments on the above magnetic resonance conveyed that the magnitude of frequency fluctuations of the Larmor precession is too small to be taken seriously as a nonlinear correction. Nevertheless, it should be noted that these experiments do not deny as yet the possibility of a nonlinear variant of quantum mechanics[126].

Is there any **real quantum chaos** that shows extreme sensitivity to initial conditions just like classical chaos? The canonical quantization method, an alternative framework of quantum mechanics, introduces the postulate of the noncommutability between canonical conjugate variables. Although this postulate is independent of the Einstein–Brillouin–Keller (EBK) torus quantization rule (see Section 5.1), the former is more or less affected by the latter. In more general cases that the genesis of chaos breaks tori, how can we justify the canonical quantization rule? The analogous question on the way of quantizing the system with no tori was raised by Einstein in 1917 prior to the birth of quantum mechanics, and it still continues to pierce through the present formalism of quantum mechanics. Feynman's path-integral method, the third scheme of quantum mechanics, is independent of both wave-mechanics and the canonical quantization method. This path-integral method has been the basis of our semiclassical theory used throughout this book. Feynman's method treats time and space on the same footing, and has the advantage of elucidating the origin of quantum fluctuations and their magnitude, so that it will serve as a powerful tool in our conceiving of a framework of nonlinear quantum mechanics that exhibits really chaotic behavior. There is the possibility of innovation of quantum mechanics in the course of exploring quantum manifestation of chaos. While nanotechnology will generate novel materials like high T_c superconductors and new devices like an ultrahigh speed quantum computer, we should also attempt to exploit this technology to verify or to *innovate* the foundations of the fundamental theory where we are peacefully living.

It is expected that, during the 21-st century, both nonlinear science and nanotechnology will break up the hard shell of fundamental science and grow up to the level that they will enrich our life and environment. In fact the study of chaos and complex systems developed so far as a branch of fundamental physics is now progressing in every direction of applied science. On the other hand, the semiconductor microfabrication technique cultivated as a mean to study fundamental properties of materials has matured and begun to be applied for many practical purposes. Considering these circumstances, the subject of *quantum chaos and quantum dots*, which bridges between nonlinear science and nanotechnology, constitutes a theme full of bright future. We hope this textbook has been very beneficial to those people entering into such a romantic and interdisciplinary field and also to those who wish to specialize in the fundamental and theoretical aspects of the field.

REFERENCES

In this textbook, except where inevitable, we do not quote references each time they appear. Further, since it is difficult to cite all references of quantum chaos exhaustively, we list only essential ones related to quantum dots or quantum billiards, which are given below almost in order of appearing in this book.

· Standard textbooks that treat chaos and ergodicity in Hamiltonian dynamical systems[1,2]

[1] V. I. Arnold: Mathematical Methods of Classical Mechanics (Nauka, Moscow, 1974).

[2] A. J. Lichtenberg and M. A. Lieberman: Regular and Stochastic Motion (Springer, Berlin, 1983).

· Standard textbooks for path-integral methods[3,4]

[3] R. P. Feynman and A. R. Hibbs: Quantum Mechanics and Path Integrals (McGraw-Hill, New York, 1965).

[4] L. S. Schulman: Techniques and Applications of Path Integration (John Wiley and Sons, New York, 1981).

· Textbooks for semiclassical theory of chaotic systems[5-7]

[5] M. C. Gutzwiller: Chaos in Classical and Quantum Mechanics (Springer, Berlin, 1990).

[6] M. Brack and R. K. Bhaduri: Semiclassical Physics (Addison Wesley, New York, 1997).

[7] P. Cvitanović et al.: Classical and Quantum Chaos: A Cyclist Treatise (http://www.nbi.dk/ChaosBook, Copenhagen, 1997).

· Early representative papers on semiclassical theory[8-16]

[8] M. C. Gutzwiller: J. Math. Phys. **8** (1967) 1979.

[9] M. C. Gutzwiller: J. Math. Phys. **10** (1969) 1004; **11** (1970) 1791.

[10] M. C. Gutzwiller: J. Math. Phys. **12** (1971) 343.

[11] M. V. Berry and K. E. Mount: Rep. Prog. Phys. **35**(1972) 315.

[12] M. V. Berry and M. Tabor: J. Phys. **A10** (1977) 371.

[13] R. Balian and C. Bloch: Ann. Phys. **60** (1970) 401 .

[14] R. Balian and C. Bloch: Ann. Phys. **69** (1972) 76.

[15] R. Balian and C. Bloch: Ann. Phys. **85** (1974) 514 .

[16] V. M. Strutinsky and A. G. Magner: Sov. J. Part. Nucl. **7** (1976) 138.

· Books with a collection of important works[17-20]

[17] M. J. Giannoni et al. (eds.): Chaos and Quantum Physics (Session LII, Les Houches, 1989) (Elsevier, Amsterdam, 1991).

[18] B. V. Chirikov and G. Casati(eds.): Quantum Chaos——Order and Disorder (Cambridge University Press, Cambridge, 1995).

[19] I. V. Leerner et al. (eds.): Supersymmetry and Trace Formula ——Chaos and Disorder (Kluwer Academic/Plenum Publishers, New York, 1999).

[20] G. Casati *et al.* (eds.): New Directions in Quantum Chaos, Proceedings of International School of Physics ≪Enrico Fermi≫ (IOS press, Amsterdam, 2000).

· Textbooks for other areas of quantum chaos[21-23]

[21] F. Haake: Quantum Signatures of Chaos (Springer, Berlin, 1991).

[22] K. Nakamura: Quantum Chaos——A New Paradigm of Nonlinear Dynamics (Cambridge University Press, Cambridge, 1993).

[23] H-J. Stöckmann: Quantum Chaos——An Introduction(Cambridge University Press, Cambridge, 1999).

· Textbooks for a border between quantum chaos and quantum transport[24-27]

[24] N. E. Hurt: Quantum Chaos and Mesoscopic Systems——Mathematical Methods in the Quantum Signatures of Chaos (Kluwer Academic, Dordrecht, 1997).

[25] K. Nakamura: Quantum versus Chaos——Questions Emerging from Mesoscopic Cosmos (Kluwer Academic, Dordrecht, 1997).

[26] K. B. Efetov: Supersymmetry in Disorder and Chaos (Cambridge University Press, Cambridge, 1997).

[27] K. Richter: Semiclassical Theory of Mesoscopic Quantum Systems (Springer, Berlin, 2000).

· Proceedings and special issue on the border between quantum chaos and quantum transport[28-30]

[28] E. Akkermans *et al.* (eds.): Mesoscopic Quantum Physics, Proceedings of Les Houches Summer School (Elsevier, Amsterdam, 1995).

[29] K. Nakamura (ed.): Special issue on Chaos and Quantum Transport in Mesoscopic Cosmos, Chaos, Solitons and Fractals **8** (1997) No.7&8.

[30] K. F. Berggren and S. Åberg (eds.): Quantum Chaos Y2K, Proceedings of Nobel Symposium (Royal Swedish Academy of Sciences/World Scientific, Singapore, 2001).

· Classical scattering theory on billiards[31-36]

[31] For the article based on this lecture, see, L. Kelvin: Phil. Mag. **Ser. 6** (1901) 1.

[32] N. S. Krylov: Works on the Foundations of Statistical Physics (Princeton University Press, Princeton, 1979).

[33] V. I. Arnold and A. Avez: Theorie Ergodique des Systems Dynamiques (Gauthier–Villars, Paris, 1968).

[34] Ya. G. Sinai: Russ. Math. Surv. **25** (1970) 137.

[35] Ya. G. Sinai: Introduction to Ergodic Theory (Princeton University Press, Princeton, 1977).

[36] L. A. Bunimovich and Ya. G. Sinai: Commun. Math. Phys. **78** (1980) 247, 479.

· Early representative experiments on quantum transport in quantum dots[37-41] and a collection of review articles on the recent experiments[42]

[37] C. M. Marcus *et al.*: Phys. Rev. Lett. **69** (1992) 506.

[38] C. M. Marcus *et al.*: Chaos **3** (1993) 643.

[39] A. M. Chang *et al.*: Phys. Rev. Lett. **73** (1994) 2111.

[40] A. M. Chang: Chaos, Solitons and Fractals **8** (1997) 1281.

[41] J. P. Bird *et al.*: Phys. Rev. **B52** (1995) R14336; Chaos, Solitons and Fractals **8** (1997) 1299.

[42] J. P. Bird (ed.): Electron Transport in Quantum Dots (Kluwer Academic, Dordrecht, 2003).

· Fundamental formulas for mesoscopic quantum transport[43, 44]

[43] R. Landauer: IBM J. Rev. Dev.**1** (1957) 223; Philos. Mag. **21** (1970) 863.

[44] D. Fisher and P. A. Lee: Phys. Rev. B **23** (1981) 6851.

· Early representative semiclassical theories on mageneto-resistance in quantum dots[45-47]

[45] R. A. Jalabert, H. U. Baranger, and A. D. Stone: Phys. Rev. Lett. **65** (1990) 2442.

[46] H. U. Baranger, R. A. Jalabert, and A. D. Stone: Phys. Rev. Lett. **70** (1993a) 3876.

[47] H. U. Baranger, R. A. Jalabert, and A. D. Stone: Chaos **3** (1993b) 665.

· Classical, semiclassical and quantum theory of billiard systems[48, 49]

[48] P. Gaspard: Chaos, Scattering and Statistical Mechanics (Cambridge University Press, Cambridge, 1998).

[49] P. Gaspard and S. A. Rice: J. Chem. Phys. **90** (1989) 2225, 2242, 2255.

· Thermodynamic formalism of chaos with repellers[50, 51]

[50] H. Kantz and P. Grassberger: Physica **D17** (1985) 75.

[51] J. P. Eckmann and D. Ruelle: Rev. Mod. Phys. **57** (1985) 617.

· Handbook of special functions including Hankel function[52]

[52] M. Abramobitz and I. A. Stegun (eds.): Handbook of Mathematical Functions (U.S. National Bureau of Standards,Wasington, 1964).

· Multiple-scattering expansion of semiclassical Green function for chaotic billiards[53]

[53] T. Harayama, A. Shudo and S. Tasaki: Nonlinearity **12** (1999) 1113.

· Extension of Gutzwiller's trace formula for: systems with continuous symmetry[54, 55]; systems showing orbit bifurcations[56-58]

[54] S. C. Creagh and R. G. Littlejohn: Phys. Rev. **A44** (1991) 836.

[55] S. C. Creagh and R. G. Littlejohn: J. Phys. **A25** (1992) 1643.

[56] M. Sieber: J. Phys. **A29** (1996) 4715.

[57] M. Sieber: J. Phys. **A30** (1997) 4563.

[58] H. Schomerus and M. Sieber: J. Phys. **A30** (1997) 4537.

· Semiclassical theory of conductivity[59, 60]

[59] G. Hackenbroich and F. von Oppen: Z. Phys. **B97** (1995) 157.

[60] K. Richter, D. Ullmo and A. Jalabert: Phys. Rep. **276** (1996) 1.

Textbooks for mesoscopic transport in general: theory[61, 62]; experiment+quantum theory[63, 64]

[61] Y. Imry: Introduction to Mesoscopic Physics (Oxford University Press, New York, 1997).

[62] C. W. J. Beenakker: Rev. Mod. Phys. **69** (1997) 731.

[63] S. Datta: Electronic Transport in Mesoscopic System (Cambridge University Press, Cambridge, 1995).

[64] D. K. Ferry and S. M. Goodnick: Transport in Nanostructures (Cambridge University Press, Cambridge, 1997).

Experiment of diamagnetism and persistent current in quantum dots[65, 66]

[65] L. P. Lévy *et al.*: Physica **B189** (1993) 204.

[66] D. Mailly *et al.*: Phys. Rev. Lett. **70** (1993) 2020.

· Early theory of diamagnetism in quantum and/or classical billiards[67-69]

[67] H. J. van Leeuwen: J. Phys. (Paris) **2** (1921) 361.

[68] M. Robnik and M. V. Berry: J. Phys. **A18** (1985) 1361.

[69] K. Nakamura and H. Thomas: Phys. Rev. Lett. **61** (1988) 247.

· Semiclassical theory of diamagnetism and persistent current in quantum dots[70-77]

[70] M. V. Berry and J. P. Keating: J. Phys. **A27** (1994) 6167.

[71] F. von Oppen and E. K. Riedel: Phys. Rev. **B48** (1993) 9170.

[72] F. von Oppen: Phys. Rev. **B50** (1994) 17151.

[73] D. Ullmo *et al.*: Phys. Rev. Lett. **74** (1995) 383.

[74] R. A. Jalabert *et al.*: Surf. Sci. **361/362** (1996) 700.

[75] S. Kawabata: Phys. Rev. **B59** (1999) 12256.

[76] J. Ma and K. Nakamura: Phys. Rev. **B60** (1999) 10676.

[77] J. Ma and K. Nakamura: Phys. Rev. **B60** (1999) 11611.

· Theory of Drude conductivity of Sinai billiard[78]

[78] R. B. Laughlin: Nucl. Phys. **B3** (Proc. Suppl.) (1987) 213.

· Critique against semiclassical theory of ballistic weak localization[79-82]

[79] N. Argaman: Phys.Rev. **B53** (1996) 7035.

[80] I. L. Aleiner and A. I. Larkin: Phys. Rev. **B54** (1996) 14423; Phys. Rev. E **55** (1997) 1243; Chaos, Solitons and Fractals **8** (1997) 1179.

[81] A. D. Stone: in Mesoscopic Quantum Physics, Proceedings of Les Houches Summer School (Elsevier, Amsterdam, 1995) edited by E. Akkermans *et al.*

[82] R. Akis *et al.*: Phys. Rev. **B60** (1999) 2680.

· New directions of ballistic weak localization, AAS oscillation and conductance fluctuations[83-86]

[83] S. Kawabata and K. Nakamura: J. Phys. Soc. Jpn. **65** (1996) 3708; Phys. Rev. B **57** (1998) 6282.

[84] Y. Takane and K. Nakamura: J. Phys. Soc. Jpn. **66** (1997) 2977.

[85] Y. Takane and K. Nakamura: J. Phys. Soc. Jpn. **67** (1998) 397.

[86] K. Richter and M. Sieber: Phys. Rev. Lett. **89** (2002) 206801-1.

· AAS oscillation in a diffusive regime[87]

[87] B. L. Altshuler, A. G. Aronov and B. Z. Spivak: JETP Lett. **33** (1981) 94.

· Self-similar conductance fluctuations[88-95]

[88] R. P. Taylor *et al.*: Phys. Rev. Lett. **78** (1997) 1952; Phys. Rev. B **56** (1997) R12733.

[89] A. S. Sachrajda *et al.*: Phys. Rev. Lett. **80** (1998) 1948.

[90] A. P. Micholich *et al.*: Europhys. Lett. **49** (2000) 417.

[91] A. P. Micholich *et al.*: Phys. Rev. Lett. **87** (2001) 036802.

[92] R. Ketzmerik: Phys. Rev. **B54** (1996) 10841.

[93] A. Budiyono and K. Nakamura: J. Phys. Soc. Jpn. **71** (2002) 2090; Chaos, Solitons and Fractals **17** (2003) 89.

[94] M. Brack: Found. Phys. **31** (2001) 209.

[95] A. Budiyono and K. Nakamura: Phys. Rev. **B68** (2003) 121304(R).

· Experiments on magneto-resistance in antidot lattices[96-98]

[96] D. Weiss *et al.*: Phys. Rev. Lett. **66** (1991) 2790.

[97] D. Weiss *et al.*: Phys. Rev. Lett. **70** (1993) 4118.

[98] D. Weiss *et al.*: Chaos, Solitons and Fractals **8** (1997) 1337.

· New dirctions in magneto-resistance in antidot lattices[99-101]

[99] F. Nihei *et al.*: Phys. Rev. **B51** (1995) 4649.

[100] T. Ando *et al.*: Chaos, Solitons and Fractals **8** (1997) 1057; J. J. Appl. Phys. **38** (1999) 308.

[101] J. Ma and K. Nakamura: Phys. Rev. **B62** (2000) 13552.

· Arnold diffusion[102-104]

[102] B. V. Chirikov: Phys. Rep. **52** (1979) 263.

[103] G. M. Zaslavsky, R. Z. Sagdeev, D. A. Usikov, and A. A. Chernikov: Weak Chaos and Quasi-Regular Patterns (Cambridge University Press, Cambridge, 1991).

[104] J. Ma, R.-K. Yuan and K. Nakamura: Commun. Theor. Phys. (Beijing, China) **39** (2003) 279; J. Ma and K. Nakamura: LANL preprint cond-mat/0108276.

· Coulomb blockade[105-114]

[105] C. W. J. Beenakker: Phys. Rev. **B44** (1991) 1646.

[106] S. Tarucha *et al.*: Phys. Rev. Lett. **77** (1996) 3613.

[107] L. P. Kouwenhoven *et al.*: Science **278** (1997) 1788.

[108] A. M. Chang *et al.*: Phys. Rev. Lett. **76** (1996) 1695.

[109] J. A. Folk *et al.*: Phys. Rev. Lett. **76** (1996) 1699.

[110] S. R. Patel *et al.*: Phys. Rev. Lett. **81** (1998) 5900.

[111] Y. Alhassid: Rev. Mod. Phys. **72** (2000) 895.

[112] E. E. Narimanov *et al.*: Phys. Rev. Lett. **83** (1999) 2640; Phys. Rev. **E63** (2001) 056204.

[113] E. B. Bogomolny and D. C. Rouben: Europhys. Lett. **43** (1998) 111.

[114] A. Budiyono and K. Nakamura: Phys. Rev. **B67** (2003) 245321.

· Textbooks for level statistics[115, 116]

[115] C. E. Porter, (ed.): Statistical Theories of Spectra: Fluctuations (Academic, New York, 1965).

[116] M. L. Mehta: Random Matrices and Statistical Theory of Energy Levels (Academic, New York, 1967); Random Matrices (Academic, New York, 1991).

Energy diffusion and nonadiabatic transition[117-120]

[117] D. A. Hill and J. A. Wheeler: Phys. Rev. **89** (1952) 1102.

[118] A. Bulgac, G. D. Dang and D. Kusnezov: Chaos, Solitons and Fractals **8** (1997) 1149.

[119] M. Wilkinson: J. Phys. **A20** (1987) 2415; **A21** (1988) 4021.

[120] M. Wilkinson and E. J. Austin: J. Phys. **A28** (1995) 2277.

Semiclassical treatment of diffraction in billiards with a flux line[121, 122]

[121] M. Sieber: Phys. Rev. **E60** (1999) 3982.

[122] E. Bogomolny *et al.*: LANL preprint chao-dyn/9910037.

References related to future prospects[123-126]

[123] S. Weinberg: Phys. Rev. Lett. **62** (1989) 485.

[124] S. Weinberg: Ann. Phys. (N.Y.) **194** (1989) 336.

[125] J. Polchinski: Phys. Rev. Lett. **66** (1991) 397.

[126] F. Laloë: Am. J. Phys. **69** (2001) 655.

INDEX